The OLDER PERSON'S *Guide to* NEW STUFF

Also by Mark Leigh for Robinson
How to Talk Teen
www.mark-leigh.com

The OLDER PERSON'S Guide to NEW STUFF

From Android to Zoella, a complete guide to the modern world for the easily perplexed

MARK LEIGH

ROBINSON

First published in Great Britain in 2019 by Robinson

All images supplied by Shutterstock

10 9 8 7 6 5 4 3 2 1

A CIP catalogue record for this book is available from the British Library

ISBN 978-1-47214-237-5

Typeset in Sentinel
Designed by Thextension
Printed and bound in Great Britain by Clays Ltd, Elcograf S.p.A.

Papers used by Robinson are from well-managed forests and other responsible sources

Robinson
An imprint of
Little, Brown Book Group
Carmelite House
50 Victoria Embankment
London EC4Y 0DZ

An Hachette UK Company
www.hachette.co.uk
www.littlebrown.co.uk

To my dad,

a man who thinks that a 'cursor' is someone with
a potty mouth and a 'bar code' is no jeans or trainers

Give a person a fish and you feed them for a day;
teach that person to use the internet and they won't
bother you for weeks.

Anonymous

Contents

Introduction

The inspiration for this book actually came from my ninety-two-year-old father who dropped a bombshell when he asked me, apropos of nothing, 'What's the internet?'

My first thought was, *Wow. Where did that come from?*

My second was, *How do I explain this to someone who has never used a computer, who thinks Facebook is a book and who views a smartphone almost as witchcraft?*

I can't remember my response at the time, but it seemed to quench his thirst for this specific knowledge. But that got me thinking: I wondered if there was a handy, easy-to-understand guide for the older (and the not-so-old) person who's genuinely baffled and bewildered by today's technology.

If there was such a thing, I couldn't find it.

So, in the absence of the book I had in mind, I decided to write it.

Aimed at people such as my father – anyone who says, 'The Twitter' or who asks, 'Do they deliver emails on Sunday?' – this book seeks to explain a host of modern concepts that have entered everyday use and parlance, but which are alien (and possibly frightening) not just to the elderly but, if truth be known, to anyone over forty-five.

As you'd expect, most of the entries are terms relating to technology – for example, *firewall, vlog* and *streaming stick* – but I've also included a number of popular modern concepts or conceits such as *crowdfunding, artisanal* or the *Kardashians,* all of which can be just as mystifying.

Where relevant, terms are cross-referenced. This can be recognised by words or phrases that appear in **bold** (although the one word that's not cross-referenced is 'internet' as it appears in about 70 per cent of entries and the publisher said they'd run out of ink if they had to use bold type for that word each time).

And one last thing: all the information and statistics are correct at the time of going to press but since technology moves at such a rapid pace, by the time this book's on the shelves 'what's hot' might well be 'what's not'.

So sit down, kick off your shoes (or comfy slippers) and get ready to learn why *spam* isn't tasty, why a *dongle* isn't as rude as it sounds or why *sticky content* isn't as gross as it might seem.

Mark Leigh

PS I've tried to give a basic overview of each term or phrase so please don't expect a long, detailed, exhaustive explanation. If you really want to find out more about something like *logic bombs* or *nanobots*, for example, then you know what to do.

Google it.

From Ad blocker *to* Avocado toast

Ad blocker

A **program** that does what it says on the tin . . . it removes different kinds of ads from **websites** (e.g. **pop-ups**) so you can enjoy web pages without any annoying distractions or interruptions. Not only that, but without having to show ads, web pages can load much more quickly.

AI

The phrase 'Artificial Intelligence' sounds quite ominous, so some people prefer to call it 'machine learning' which sounds less threatening – until, that is, you think about the *Terminator* movies.

Putting aside the scenario of a nearly extinct human race battling the marauding forces of Skynet for a moment, AI is a branch of computer science that refers to an intelligence learned and demonstrated by machines as opposed to the natural intelligence displayed by humans and animals. A more detailed definition of what constitutes AI is any system or **program** that demonstrates some of the following traits of human intelligence: perception, learning, reasoning, planning, problem solving and, to a lesser degree, social intelligence and creativity.

AI sounds like something very futuristic but the truth is, we're surrounded by AI technology today that we all take for granted. This includes computer programs that understand and respond to human speech (like **Alexa** or **Siri**) or those which can analyze familiar faces in a photograph (like **Facebook**). Electric toothbrushes that give feedback to users about the correct way to brush and robotic vacuum cleaners that learn the layout of a room are also examples of AI – as are **search engines**, **spam** email filters and the plagiarism detectors used by schools and colleges to check students' work.

Over the next few years, providing we don't all perish at the hands of an evil sentient computer, the development and adoption of AI will be one of the most exciting and game-changing aspects of technology. This includes the more widespread use of **autonomous cars**, medical tools that can identify malignant tumors in seconds or robotic limbs or exoskeletons to help those unable to walk.

Those worried about AI claim that it's difficult to think of a job that a computer won't be able to do as well as, if not better than, a human – and that poses a whole series of profound challenges for society.

Wow! That's quite heavy and it's only the second entry in this book ...

Airbnb

Technically this is an **online** hospitality marketplace, but you probably know it as 'that room-renting **website**'. Airbnb and similar companies provide a way for people who *need* accommodation to make contact with people who are *offering* accommodation. The accommodation might be a single room or a whole home – and the home could be an apartment, a house, a villa, a holiday cottage, even a castle – anywhere in the world. These are short-term rentals, from one night upwards; the maximum length of time you can stay is determined by the host or local legislation.

After you register on the Airbnb website, you select what you're looking for in terms of the location and your

budget and dates, then Airbnb presents you with a list of properties that meet your criteria.

Airbnb makes its money by acting as a sort of broker; it charges a percentage fee on every booking from both the guest and the host. The company has over 4 million listings in 65,000 cities in 191 countries.

Unusual Airbnb accommodation has included a lighthouse, a treehouse, a shipping container, a yacht, a covered wagon, a windmill, a Romany caravan, an igloo and a converted Edwardian railway carriage.

Alexa

This is **Amazon**'s **digital assistant**, voice-recognition **software** built into its range of Echo products, and the equivalent of **Apple**'s **Siri**. She responds to the **wake words**, 'Alexa', 'Amazon', 'Echo' or 'Computer', and you can interact with Alexa, asking her questions or giving commands as if she was a real person. Alexa will respond by voice and/or carry out your requests. The sort of things Alexa can respond to are, 'Will it rain next Tuesday?', 'Wake me up at 6 a.m. tomorrow', or 'Play me songs by Jimmy Osmond'.

Don't worry about this last command. Alexa is very non-judgemental.

Algorithm

This is a computer term used by people who want to appear extra clever. It sounds very technical, but an algorithm is just a set of instructions or a process for performing a task. A recipe or a set of directions to your house is actually an algorithm . . . it's just that no one uses the term in this way. (Ask your Aunty Gladys for the algorithm for her lemon drizzle cake and she'll give you a very peculiar look.)

The word algorithm is normally used when describing a sequence of step-by-step instructions in a computer **program** that tell the computer what to do. For example, a **search engine** algorithm is what's used to display a list of the **websites** that match what you're looking for.

And that's really all you need to know about the subject.

Amazon

If you bought this book via Amazon then you can probably skip this entry (just remember to leave a five-star review). If you're not one of the 310+ million active Amazon buyers then read on.

That statistic alone gives you an idea of the scale of what is the world's largest internet retailer measured by revenue – a company that now employs nearly 566,000 people worldwide on a full and part-time basis. Founded in 1994 and operating from its owner's garage, Amazon started by just selling books, then diversified into videos, electronics, clothes, furniture, food, toys, jewellery – in fact, everything!

See also: **Alexa**, **Fire TV Stick**, **Kindle**

Eight things you may not know about Amazon

1 Amazon's founder, Jeff Bezos, was originally going to call Amazon 'Cadabra' (as in 'Abracadabra') but changed his mind after his lawyer misheard it as 'cadaver'. The second name was 'Relentless' but this was dropped after friends told him it sounded a bit sinister.

2 The first book Amazon sold was Douglas Hofstadter's *Fluid Concepts and Creative Analogies: Computer Models of the Fundamental Mechanisms of Thought.*

3 The person who bought it, John Wainwright, had an Amazon building named after him to honour the occasion.

4 In 1997 US bookstore Barnes & Noble sued Amazon alleging that its claim that it was 'Earth's largest bookstore' was false as it was actually a book broker rather than a bookstore. The lawsuit was settled out of court and Amazon continued to make that claim.

5 Amazon invented the concept of '1-click buying' and licensed it to **Apple** for its **iTunes** purchases.

6 In the early days of Amazon, Bezos would award a prize to staff members who found the strangest book titles each week. Reported winners included *How to Start Your Own Country* and *How to Train Goldfish Using Dolphin Training Techniques* (both of which are still available on Amazon).

7 One of its fastest ever deliveries was an Amazon Echo which, in July 2017, arrived at its destination in Sutton Coldfield in the West Midlands just fourteen minutes after being ordered.

8 In 2012, Amazon's **website** went down for forty-nine minutes. As a result, the company estimated it lost nearly $5.7 million in sales.

Android

In this sense, it's not a robot with a human appearance but the world's most popular **operating system** for **smartphones**, **tablets** or **smartwatches**. Developed by **Google**, Android is the main rival to **Apple**'s **iOS** system. There are a few other operating systems for mobile devices including **BlackBerry** OS, Symbian (used by Nokia), **Windows** Phone OS and Windows 8 but these are in decline faster than **SCART leads**. Last year Android and iOS accounted for approximately 99 per cent of all new phone and tablet operating systems, but is Android better? Well, it's on a wider range of phones and they tend to be more affordable, plus you get more **Cloud** storage as standard. Apple's iOS has other benefits so there's no clear winner. Most people choose their phone first and live with whatever operating system it comes with.

Animated GIF

The Mona Lisa winking . . . Einstein poking his tongue in and out . . . the 'Dancing Baby' from *Ally McBeal* in the late 1990s . . . these are all examples of animated GIFs – often found on **websites**, in **emails** and shared across **social media** (the Dancing Baby became one of the first **viral** videos). As the name implies, this is a short animation (it can be words as well as images) that may be looped or end after a few seconds. The truth is, it's actually just a series of static images in quick succession that give the impression of movement.

You can make your own animated GIFs using **Photoshop** or a number of specialist **software** packages such as Giphy (www.giphy.com) or Gifmaker (www.gifmaker.org).

Anime

Pronounced 'an-i-may', this is a very popular form of Japanese film and TV animation and, while you might not know the name, you'll probably recognise the style. Whereas cartoons are more comical and primarily aimed at children, anime tends to feature more serious story-telling and is produced for all ages. This is what makes anime different from conventional cartoons:

* Anime characters are more realistic, although many have disproportionally large eyes and small mouths, which have the effect of creating a 'cuter' face.
* Brighter colours.
* Adult content (I'm not talking about 'sexy time', but more elaborate plots about things such as love, loss and conflict – and lots of emotion).
* Longer: approximately twenty-two to twenty-five minutes per episode with anime movies being considerably longer.

See also: **manga**

In Japan, the word anime is used to refer to all animation. Outside of Japan, it's become the catch-all term to describe animation specifically from Japan.

App

App is short for 'application', which in turn is short for 'software application' – in other words an app is a type of **software program**; one that's been designed to work specifically on **mobile devices**. The tube map on your phone? That's an app. So too is **Facebook**, **Instagram** or **Twitter**. And that calorie counter, the foreign currency convertor, the game where you have to run over zombies, the restaurant finder, the taxi booker or that thing where you can swap your face with that of your dog – these are all apps. Apps are either free or paid for, but are designed to work with specific **operating systems** so you need to **download** them from an **app store** relevant to your device. There are over 6 million apps available covering probably anything you'd want to do – and some that you probably never will. For example, S.M.T.H. – or Send Me To Heaven – is an app that records how high you can throw your **Android** phone and **uploads** results to a list of rankings. **Apple** refuses to distribute this app for its phones on the grounds that it was 'encouraging behaviour that could result in damage to the user's device'.

Killjoys.

Apple

One of the best-known technology companies and the creators of products including the **iPad**, **iPhone** and **iPod**, the MacBook range of laptops, the Apple Watch and **Apple TV**, and services including **iTunes** and **Apple Pay**. Co-founded in 1976 by the late Steve Jobs, Apple is a company that polarises people.

Its critics claim that it doesn't innovate, but just takes existing products and improves them (which, it must be said, is not in itself a bad thing), and that it has a policy of bringing out new versions of its products at an alarming rate, making models just a few months old obsolete (or, if not obsolete, then terribly unfashionable) and that its products are exceedingly overhyped and overpriced.

Fans of Apple, however, point out that its products are not just technologically advanced but are designed to be intuitive and easy to use – and that they look sooooooo cool. They also point out Apple's reputation for offering a great customer service and in-store experience.

Steve Jobs named the company after returning from an apple farm. His biography claims he found the name, 'fun, spirited and not intimidating'.

Apple Pay

Another step in the inevitable march towards a cashless society, Apple Pay is a contactless payment system that works with the **iPhone** and Apple Watch – the advantage being that it works above the current £30 limit for contactless credit and debit cards (although the actual limit depends on the retailer).

Advantages

- Easy to use
- Private and secure
- Higher payment limit than contactless cards

Disadvantage

- The convenience makes it all-too-easy to spend £3.80 a day on an Iced Cinnamon Almondmilk Macchiato

See also: **Apple**

Apple TV

Not an actual TV *per se*, but **Apple**'s own **set-top box** that allows the user to receive digital **content** (TV programmes, films, music), as well as internet access, and play it on a connected TV. What makes Apple TV different is the fact that you can also use it to play content directly from your **iPhone**, **iPad** or Macbook on to a TV as long as the devices are all on the same **WiFi network**. The latest version also uses **Siri** to react to voice commands.

App store

Any place where you can search for, and **download apps** is an app store. Although **Apple**'s app store is called App Store, it's not an Apple trademark (but it did unsuccessfully try and make it only apply to its own apps). Other major app stores are the **Amazon** Appstore and **Google Play** (both for **Android** devices) and the **Microsoft** Store (for **devices** running **Windows**).

Artisanal

Long, long ago the adjective artisanal, the art of creating something by hand through a combination of long practice, dedication, aptitude and skill, was used to describe crafts such as carpentry, stonemasonry and decorative ironwork.

Now it's overused and over-hyped, misappropriated by businesses catering for **hipsters** and **millennials** to describe things such as cheese, pickles, coffee, bread – and even water. Yes, there is such a thing as artisanal water. There's no actual legal definition of what constitutes an artisanal product and it's become less of a valid way to describe a production process and more of a trendy marketing term (no doubt touted about by artisanal marketing agencies).

Attachment

This is the 'catch-all' name given to any separate **file** that's sent in conjunction with an **email**. It can be a photograph, a document, a **PDF**, a movie, music – or any other type of file that needs a separate **program** to open it.

Attachments can sometimes contain **viruses** or **malware**. To protect your computer or **mobile device** it's safer not to open any attachment you weren't expecting.

Augmented reality

Unlike **virtual reality** where you inhabit a simulated 'virtual' world, augmented reality (known as AR) provides an enhanced view of the *actual* world – by combining images of your real environment with superimposed computer-generated images or text. The result is a heightened view of reality that's become very popular in smartphone apps and video games.

Google Translate is an example of an **app** that features AR technology; you can point your phone's camera at a menu for example and the text is translated into a language of your choice – the words being superimposed on the actual menu. Other AR apps include those that overlay constellations and planets over any part of the night sky, allow you to see what you'd look like with make-up, a new hairstyle or clothes, view different types of furniture in a room, or superimpose a proposed navigation route over a view of a real street.

There's even an app (Ink Hunter) that lets you view what your tattoo will look like on any part of your body, painlessly.

See also: **Pokémon GO**

Autonomous car

A science-fiction film cliché, an autonomous car – also known as a driverless car or self-driving car – is a vehicle that can sense its environment and guide itself without any human interaction. This type of vehicle is currently being tested on public roads in various cities around the world, in particular by **Uber**, which plans to launch a fleet of driverless taxis. Despite this ongoing development and testing, no autonomous cars are currently in use commercially; several have been involved in crashes and the death of a pedestrian.

They use a combination of **GPS**, cameras and LIDAR (a new type of radar that uses light rather than radio waves) to navigate and drive safely. There is, however, a huge number of technical obstacles to overcome before they can be widely adopted. These include the ability to recognise hand signals from traffic police, dealing with traffic lights that are broken, snow or heavy rain confusing their radar, unclear lane markings – and, of course, the actions of unpredictable human drivers with whom they're sharing the roads.

Auto-Tune

Think of this as a lifebelt that producers can throw to singers floundering in a sea of flat notes and tone deafness. A piece of proprietary **software** that's pretty much standard in all professional recording studios, Auto-Tune

has become a generic term for disguising or correcting off-key vocal performances; the technical term for this process is 'digital pitch correction'. After Auto-Tune works its magic, the result is a vocal track that's perfectly in tune despite the singer's inability to hit the right note, let alone hold it.

According to Antares Audio Technology, the company that holds the patent for Auto-Tune, it's used on almost all vocal performances you hear on the radio; in 2010 the British version of *The X Factor* admitted to using Auto-Tune to improve the voices of contestants. Apart from making a singer pitch-perfect, Auto-Tune can also be used to radically change a voice for effect. It was Cher's electronic warbling voice on her huge 1998 hit 'Believe' that brought Auto-Tune to the attention of the masses.

In 2010 *Time* magazine included Auto-Tune in its list of 'The 50 Worst Inventions', calling it, 'A technology that can make bad singers sound good and really bad singers sound like robots.'

Avatar

Unlike **profile pictures**, avatars are small images (also called icons) used to represent a person in a video game, on **social media**, an **email** account, **online forum** or **chat room**. Avatars can be anything: caricatures, logos, graphic images, photographs. They can be symbolic, ironic or completely arbitrary. In many cases avatars are people's fantasies of who they want to be, so next time you see someone depicted as Conan the Barbarian, it's quite probable that it's a bespectacled fifty-something accountant called Colin.

NB A profile picture can be an avatar, but an avatar isn't a profile picture.

Avocado toast

It's difficult to think that toast could ever be fashionable but spread smashed avocado on it, add lemon juice, red pepper flakes and a hint of garlic and you've got the most **on-trend** toast imaginable. Until the next on-trend toast, that is.

Served in **artisanal** coffee shops and beloved of **millennials**, it's been documented that avocado toast went from 'interesting breakfast item' to 'phenomenon' when Gwyneth Paltrow included a recipe for it in her 2013 cookbook, *It's All Good*. She commented that it was like a favourite pair of jeans: 'so reliable and easy and always just what you want'.

The popularity of the food grew with **social media**, and the sharing of recipes and photos; there are currently over 790,000 **Instagram posts** for #avocadotoast.

From Back-up *to* Buzzfeed

Back-up

Backing up is the process of copying and storing your computer data for reasons of security and safety, i.e. in case documents, photographs, music, etc., are lost due to being accidentally deleted or corrupted by a **virus**. Creating a back-up means that your data will be stored in at least two places; what's vital, however, is that those two places aren't on the same device. For example, you may feel good about having all your wedding photos and a complete copy of them on your **laptop** but if it's stolen, you lose all of them... Duh!

A sensible back-up plan means storing them separately from your computer; for example, on **flash drives**, **hard drives**, CD-ROMs, DVDs or backed-up to the **Cloud** via a service such as **Dropbox**. If the files are small enough you can also **email** them to yourself and keep them in your email account.

Bcc

You'll see this on the top of an **email** message. It stands for 'blind carbon copy', a back-in-the-day reference to when something called carbon paper was used to make a copy of a typed letter. Entering someone's email address in the Bcc box means that person will receive a copy of the message – but the actual recipient of the email (or anyone else copied into it) won't know!

Sneaky, huh?

See also: **Cc**

Bing

A **search engine** owned by **Microsoft** which has just over 3 per cent of the market share worldwide across all types of device, compared to its rival **Google** on nearly 91 per cent. Bing is pretty good, though it's been said that the name stands for Bing Is Not Google . . . that's the sort of joke software engineers like making.

Binge-watching

Bingeing is never good, whether it's cake, chocolate, pies, beer or, in this case, TV shows. The term is usually taken to mean watching four or more episodes of the same series in one sitting. Binge-watching (or binge-viewing) is considered by some to be an addiction and is a fairly recent phenomenon, driven by the introduction of DVD **box sets**, the growing popularity of **video on demand streaming** services and better-quality programming.

Biometrics

The technical term for the analysis of the way people look and act in order to identify individuals. It's the basis of what's called 'biometric authentication', which works on the basis that every single person can be identified by features that make them unique. This includes things such as their facial features, iris, voice and, of course, fingerprints – and also things such as the way they walk (called 'gait recognition'). Biometrics is used to recognise and track anyone under surveillance, but it's also used to recognise individuals in order to allow them to access a computer or **mobile device** without the need to type in a **passcode**.

See also: **faceprint**, **facial recognition**

Bitcoin

The original and one of the best-known **cryptocurrencies**, Bitcoin was created in 2009 although no one knows who invented it; the person (or group) responsible is only known by the name Satoshi Nakamoto. Bitcoin rose from obscurity to mainstream recognition thanks to what's been described as an 'insane' surge in value in 2017, and its acceptance as a form of payment by many worldwide brands. Like all cryptocurrencies, Bitcoin has an equal share of advocates as it does detractors, with high profile investors such as the CEO of JPMorgan Chase and high-profile investor Warren Buffett calling it, respectively, a 'fraud' and a 'mirage'.

One thing's certain, though: when it comes to discussing the value fluctuations of Bitcoin, the two most common words used are 'volatile unpredictability'.

See also: **blockchain**

BlackBerry

If this was 2008 and you were a thrusting young Turk in business, then the chances were you'd be carrying a BlackBerry Bold in a belt holster. At its peak, BlackBerry was one of the most prominent and innovative **smartphone** manufacturers in the world and one of the few to persist with physical keypads (up until 2017). There were once 85 million BlackBerry users in the world; that figure is now down to about 11 million as a result of competition from more technically advanced **Android** and **iOS** smartphones. In a 'if you can't beat 'em, join 'em' move, the company introduced an Android-based phone in 2015.

The name BlackBerry was chosen due to the way the tiny keyboard buttons resembled a blackberry fruit.

Blendr

The heterosexual dating **app** from the company behind **Grindr** (and still with that annoying missing 'e') but less focused towards dating and sex and more towards creating friendships between people who share the same hobbies and interests; connections that may or may not end up as relationships.

Block

To block someone on **social media** means you prevent them from seeing your **profile** or anything you **post**; that means they'll no longer be able to see what you're doing or who you're doing it with. You can do this within the privacy settings of your social media account. Blocking can be done for a number of reasons but in most cases it's to stop an ex- or someone you've fallen out with knowing your business – or to distance yourself from someone who's abusive, harassing you or just being plain creepy. Blocking someone doesn't mean it's permanent; it can

easily be undone. The person being blocked isn't notified . . . but they'll know when they realise they can't see anything you post, start a conversation with you or invite you to events, etc.
See also: **unfriend**

Blockchain

This is quite possibly the most difficult term in this book to explain and, quite frankly, you'll probably never, ever have to use it. But, for those with an unquenchable thirst for knowledge – or showing-off – read on . . .

Blockchain is the technology that underpins how a **cryptocurrency** works. A common analogy is that it's like a digital ledger; a massively huge spreadsheet available across computer **networks** that records all digital currency transactions immediately they take place. This ledger is available to anyone (via their heavily encrypted password) so that transactions can be checked and verified, and since the information exists simultaneously in multiple locations it's almost impossible to **hack**.

Why is it called blockchain? In the world of digital currencies a 'Block' is a record of a new transaction. When it's completed it's added to the chain of all transactions. So now you know.
See also: **Bitcoin**

Blog

Blogs are a sort of personal **online** diary or journal, although critics claim they're often forums for whiny, annoying, dull, smug, patronising people to present their pointless, vapid opinions.

Sometimes they are . . .

Think of a blog as your own **website** on a subject that interests you, which you update on a regular basis. It can be about anything: trainspotting, self-defence, Victorian drainpipes, the band Bucks Fizz, snakes as pets, gazebos, knitting baby clothes, devil worship, wine, barbed wire, vegan recipes, urban cycling, vegan cyclists . . . anything you want – there are even blogs about blogs. All you need is a passion (or at least a strong interest) in the subject and the time and motivation to update it regularly and interact with the people who are viewing your blog. Someone who creates and manages a blog is called a blogger, and the process of writing and maintaining a blog is known as blogging.

Blogs are easy to create with a type of **software** known as a **blogbuilder**.

So who blogs? For some it's just a hobby and way to interact with likeminded people – sharing specialist or technical knowledge about a topic. Others use their blog to spread their political or ideological views. Some companies will even add a blog to their website as another way to promote their business.

Now you might be asking, can I make money from a blog by getting businesses to advertise on it?

The answer is yes, but being realistic about things, you probably won't.

Your blog needs a substantial audience for businesses to see any value in advertising on it. That means you need to give people a real reason to visit it and revisit it, and the only way that happens is by including **content** that's worth reading.

Some blogs have grown to become businesses in their own right, attracting millions of visitors each month and generating millions of pounds in advertising revenue. The most successful include the Huffington Post (a liberal news blog now named HuffPost), TMZ (celebrity news), Gizmondo (technology) and Business Insider (business news, particularly finance).

See also: **microblogging**, **vlog**

There's often a fine line between blogs and websites. These are the main differences:

* Blogs are updated much more frequently . . . this can be daily or several times a week.
* Blogs allow for readers to comment and have a discussion, not just with you, the person who created the blog, but with others who read it too.
* Blog articles are always presented in reverse chronological order (newest ones at the top) and, to make navigation and searching easy, are arranged into groups and sub-groups.

Blogbuilder

This is **software** you need to create and manage your **blog**. Depending on the degree of sophistication you need, some are free and some require a monthly subscription – but all are relatively intuitive and simple to use. Examples of blogbuilders include Wordpress, Wix, Blogger, Squarespace, Ghost, Joomla, Weebly and **Tumblr**.

Blogosphere

The name given to the **blog** universe, i.e. all 440 million+ blogs on the internet. Hardly anyone really uses this word any more and neither should you.

Blue tick

The thing about **social media** is that it's easy to masquerade as someone else – or be mistaken for someone else. Depending on the circumstances this can be humorous or confusing – particularly if you think you're discussing the character subtleties of *Erin Brockovich* with the multiple Oscar-winning actress Julia Roberts whereas in reality you're just confusing a retired headmistress from Aberdeen who shares her name. To get around this problem, **websites** such as **Facebook** and **Twitter** offer a verification process for celebrities and household names in the fields of music, TV, film, sport, fashion, business, the media, etc. Once you're verified, you receive a blue tick next to your account name; this shows you're the real deal.

Twitter and Facebook have recently relaxed the criteria for verification; you can now apply for blue tick status if you're a 'public figure'; this doesn't mean you're necessarily famous, just that you have a public profile. The idea is to indicate a high-quality account with a large number of users/**followers**.

Bluetooth

This is an industry standard for the wireless transmission of data over short distances. What that means is that with Bluetooth you can connect **devices** or **peripherals** without the need for pesky cables that are invariably a) too short and b) tangled. Equipment you can connect this way is indicated by the Bluetooth symbol or the term

'Bluetooth-enabled', and includes headphones and **MP3 players**, printers and computers, games controllers and games consoles, etc.

Liberate your portable devices from their shackles! *Viva la revolución!*

Bluetooth is named after the tenth-century Danish King Harald Bluetooth who united disparate Danish tribes into a single kingdom. His nickname is thought to derive from the fact that he had a bad tooth that appeared conspicuously blue (but which actually translates as dark-coloured).

Blu-ray

A type of DVD that delivers higher quality images, i.e. full high-definition picture quality video and games. Blu-ray players can play 'standard' DVDs (and CDs for that matter), but DVD players cannot play Blu-ray discs.

Bill Gates, principal founder of **Microsoft**, once commented that Blu-ray would be the last disc format the world would ever see – and he may well be right. From **downloadable MP3** tracks and films to **Netflix** subscriptions, more and more music and video is being delivered digitally instead of on shiny plastic discs.

Bootcamp

You know those keen-looking yet weary and sweaty people you see jumping up and down or boxing in the park when you're taking your dog for a leisurely walk? Well, those people are at a bootcamp. Taking their inspiration from military training camps, characterised by a shouty man dishing out harsh discipline, these fitness bootcamps are often conducted by **personal trainers** or indeed, former military personnel, and concentrate on short bursts of intense training, usually over a four- to six-week period. The camaraderie of these group sessions often helps with motivation, especially for people who get easily bored in a gym.

Bounce

Emails that fail to reach their intended destination are said to have bounced, i.e. bounced back to the sender. A 'hard bounce' is where you've typed an incorrect email address or an email address that doesn't exist any more; in this case, your message will never, ever be received. A 'soft bounce', however, is where there's a temporary fault preventing your email being delivered; usually the result of the recipient's inbox being too full. In these cases, your email will probably eventually be delivered (but it's worth resending it at a later time, just to make sure).

Box set

Although it can refer to music and books, this term usually refers to DVDs or **Blu-ray** discs of TV shows and films that are packaged as a single unit. In the case of TV shows, this could be a season or an entire series. With movies, the box set could be a series of films (e.g. *The Lord of the Rings*), different versions of the same film (e.g. *Blade Runner*) or the collected works by a particular director (e.g. Alfred Hitchcock).

Video on demand streaming services often offer whole TV series, which they market as box sets. This has given rise to the phenomenon of **binge-watching**.

Broadband

Back in the day, you'd access the internet by means of a dial-up **modem** connecting your computer via your home telephone line. This method had three drawbacks: a) no one could use the phone while you were **online**, b) it was slow – actually, make that incredibly, painfully slow, and c) connections were preceded by what sounded like a random selection of annoying squawks, beeps and bongs underpinned by a thick layer of hissy static.

Broadband, on the other hand, is a rather loose term but in essence it means a high-speed internet connection that's much, much, much faster than dial-up. What's more, it's always 'on', so you don't have to actually connect to it each time you want to access the internet.

How is it faster? Well, it still uses the phone line (or fibre optic cable) but divides it into different channels to carry high-speed voice, data and video simultaneously.
See also: **broadband router**, **mobile broadband**

Even a slow broadband line today is usually about nine times faster than the best dial-up connection. A moderately fast broadband line can be over a hundred times faster.

Broadband router

A piece of **hardware** usually provided by your **ISP** (it looks like a slim box with flashing lights). Most modern routers aren't just routers, they also include an integral **modem**. It's this piece of kit that provides you with your internet connection. The router's job is to distribute (or route) this connection to any computer (or **device**) in your home via a physical **Ethernet** connection, and/or wirelessly via **WiFi**.
See also: **broadband**

Browser

Chrome, Internet Explorer, **Safari**, Firefox . . . these are all types of browsers and, although they look different, they all perform the same task – allowing you to access the internet. Clicking on these **programs** doesn't just give you access to the **World Wide Web**, browsers also perform several other important functions including the ability to store your favourite **websites**, set up your **homepage**, keep a history of websites visited, block **pop-ups** and set parental controls to stop children accessing 'unsuitable' websites.

BuzzFeed

US-based digital media company that provides news and entertainment **content**, primarily those stories that are **trending** on **social media**. This includes everything from in-depth analysis of world events to 'The fifty cutest puppy photos'.
Visit: www.buzzfeed.com

From Candy Crush *to* Cyber vetting

Candy Crush

Or to give it its full name, Candy Crush Saga, this is a multi-level, puzzle-type video game for **mobile devices**, which has been **downloaded** over 2.75 billion times. Candy Crush is mind-numbingly simple yet very addictive, with developers adding new episodes each week; there are currently over 3,000 levels. Think of it as a video game for people who don't play video games. While you can play for free, the game encourages you to progress by purchasing extra lives, new moves and new episodes.

Fans of the game state that it's very polished, accessible and playable in brief periods while critics claim (according to the *Guardian* newspaper) that the people who play Candy Crush Saga are 'stupid, easily-manipulated sheep who wouldn't know a proper game if it bit them on the nose'.

Cat 5, 6, 7

Different types of **Ethernet cable** ('Cat' is short for category) designed for high-speed communication between devices. In simple terms, the higher the number the better, as this represents a faster speed at which the data can be transmitted.

Catfish

You strike up an **online** relationship with the man (or woman) of your dreams. In this case he looks like a hunk, claims to be a dot.com entrepreneur and part-time model, holidays on Fiji and drives a black Lamborghini Aventador. You swap messages for months and months and each time he reveals another insight into the glamorous A-list world he inhabits. Then it's time to finally meet at a Nando's of his choosing. That's when you realise his photo and CV are totally fake and he's actually a photocopier salesman from Swindon with a bad case of B.O., who lives with his mum.

That's when you know you've been catfished.

A noun and a verb, a catfish is someone who pretends to be someone else over the internet, usually to deceive someone into pursuing a relationship. It's most common on **social media** or dating websites, where it's easy for the catfish to either create a complete fake identity or assume the identity of someone else for the same effect. In many cases, and particularly if the target is vulnerable or gullible, the deception can continue for months or years – sometimes extorting money in the process. The most obvious sign that you're being catfished is when your love interest refuses to meet. It's worth remembering that if he/she seems too good to be true, then there's probably a reason.

Spencer Morrill from Knoxville, Tennessee believed he was in a serious online relationship with singer Katy Perry for six years (despite her marrying Russell Brand in that period) before the truth dawned on him. 'Katy Perry' turned out to be a girl from Gloucester.

Cc

If you want to copy other people into an **email**, enter their email addresses in the box marked 'cc'. Unlike **Bcc**, this means that everyone named on the email will not only see the message, but exactly who it was sent to.

CGI

Toy Story, Avatar, The Polar Express or anything involving people in Spandex and capes . . . movies with diverse visual styles but all using computer-generated imagery to make the unreal real. Today, CGI is a mainstay of movie visual effects and computer games but when the first extended use of CGI was used in a film (*Tron* in 1984), it was viewed by some as a gimmick and by others as ground-breaking. John Lasseter, head of Pixar and Disney's animation group, stated that, 'Without *Tron,* there would be no *Toy Story*.'

Chat room

Think of this like a virtual party; a place where you can meet new people and talk to them without having to dress up or do your hair – and with no fear of your partner whinging, 'Can we go now?' or 'Stop that. You're embarrassing me.' Chat rooms are **online** social environments where users type short messages to communicate with multiple people in real-time.

Anyone in the 'room' can initiate a conversation or respond to one – and see everyone else's comments instantly (each user's response appears in a different colour). Chat rooms are places where complete strangers can **share** information or ask for advice on a particular topic, hobby or interest, argue about politics or sport – or just like real parties, flirt (plus, you can be conversing in just your underpants . . . no one need ever know).

See also: **forum**, **moderator**

Chrome

See: **Chromebook, operating system**

Chromebook

A generic term for a low-cost, lightweight **laptop** that runs **Google**'s own **operating system** called **Chrome**. Designed primarily to be used when connected to the internet, Chromebooks store **files online** in the **Cloud** rather than on the device itself. They also have a far longer battery life. If most of your computer usage is **surfing the net**, using **social media** and **emailing**, then a Chromebook is ideal.

Clickbait

This is exactly what the name implies: an eye-catching headline and image on a web page that entices you to click on it to satisfy your curiosity. Sometimes they appear under a heading such as 'Sponsored stories' or, even more dubiously, 'Recommended by the web', but in any case, this is the sort of thing that would attract your attention:

'This man tried to French kiss a tiger. You won't believe what happened next!'

'These facts about bananas will change the way you live your life forever!'

'Twenty-one ways your hairdresser can control your mind.'

Being intrigued and clicking on the headline (which is a **hyperlink**) will take you to that feature. This could be on a news **website** or a site with an innocuous and independent-sounding name such as 'Web News Update'. The intriguing headline is used in two ways. It could be a link to an article that's really just a poorly disguised advertisement, or it's a way to get someone to a website with the aim of them then exploring as many other articles/stories as possible. The more people the website attracts and the longer they stay there – the more advertising revenue they can generate.

Cloud

There are two ways to store computer **files**. The first is on physical devices such as a **hard drive**, **flash drive** or a DVD/**Blu-ray** disc. The other way is to store them remotely. I'm not talking in another room. Or in the garage. This way is a lot, lot further than that. Here, files are stored via the internet on huge remote physical **servers** that could be the other side of the world. It's because these servers are so distant – and so intangible – that this storage is known as the Cloud.

The advantage of this method is that storing data on the Cloud frees up your own storage and it lets you access your files via the internet from any device, anywhere in the world. You can also set your device to automatically **back-up** your files to the Cloud at regular intervals. Disadvantages are that some people have fears over security/privacy, and if you don't have web access, you can't access your data.

Three major Cloud service providers are **Microsoft** (OneDrive), **Apple** (iCloud) and **Google** (Google Drive). All offer a free allowance for storage and the option to pay for more if you need it.

Content

Not content as in 'satisfied' but a term that refers to the many and varied elements that can be found on a **website** or in a **post**, e.g. text, photos, videos, graphics, animations, sounds.

In other words, 'stuff'.

Cookie

You know when you're on a **website** and a **pop-up** ad suddenly appears for something you were searching for/ reading about **online** recently. Spooky, isn't it? It's as if the internet knows what you're interested in and has been following you around all sorts of different websites.

Surprise!

It has.

And it does this by cookies (also called web cookies, internet cookies or browser cookies). These are small pieces of data sent out by a website that you visit, which are then stored on your computer by your internet **browser**. Their job is to record your browsing history, what sites and topics you're interested in. Then, when you're on another website, they present a paid-for ad that they feel is relevant – whether it's for a pair of knee-high suede boots, a new rotary lawn mower or a Ukrainian bride.

Remember, companies or organisations claim that they use cookies to 'improve the user experience'. By this they mean, they 'want to sell you something'.

NB You can disable or turn off cookies on your computer or **mobile device**. Depending on the browser you're using, this function can usually be found under the 'Privacy section' under the **menus** for either Tools, Preferences or Settings.

Cosplay

A contraction of the phrase 'costume play', this is the hobby of dressing up and pretending to be a fictional character. Participants are called cosplayers and can take their inspiration from a wide variety of sources including comic books, live action films, television series, **anime** and video games. Although people have dressed up as their heroes (or favourite villains) for decades, organised

cosplay is a relatively recent phenomenon, spreading via science-fiction fan conventions. So popular is the activity, there are even cosplay-focused conventions.
See also: **fanboy/fangirl**

The first cosplay participants at a science-fiction convention were Forest Ackerman and Myrtle Douglas who attended the 1st World Science Fiction Convention in New York in 1939. Ackerman had thought everyone was going to be in costume; they were the only ones who'd dressed up . . .

Crowdfunding

Say you wake up one day, revitalised with the realisation that your life has a purpose . . . and that purpose, your mission in fact, is to make a documentary about the plight of Peruvian beekeepers. You work out that, all in, you'd need £20,000 to fund it. The question is, where do you find that money? Your credit card has a limit of £750, you're too embarrassed to ask friends and family, and your bank manager will just point at you and laugh.

The modern answer is crowdfunding, a relatively new way of raising finance from potentially millions of funders over the internet by seeking a large number of people (your backers) to each contribute a relatively small amount of money.

Crowdfunding consists of three elements:

1 The project initiator (i.e. you) who proposes the venture for which funding is needed, giving as much information as possible about it and setting a target for how much money you need.
2 Individuals or groups who donate money within a given time period.

3 A **platform** that brings 1 and 2 together (and which earns money by taking a commission on the funds raised).

If you reach your target in the set time period, you get to keep the funding. If you don't, then no money is taken from the backers.

So, in the case of your beekeeping documentary you just need 20,000 people to each donate on average £1 to your project; or 50p from 40,000 people . . . a much more manageable and practical ask.

So why should someone back someone they don't know and will probably never meet? Well, for some it's just a feeling of philanthropy, to help someone realise their dream or the chance to support a cause they believe in. For others, it could be a bit of that plus the fact that they'll be in line to earn tangible rewards for their funding, e.g. a free T-shirt or DVD, or an invitation to the premiere.

Crowdfunding is often used for more creative projects such as producing films, music, plays, stage shows, publishing, fashion, video games or new technology. Some of the best known crowdfunding organisations are **Kickstarter**, Indiegogo, Patreon, Rockethub, Seedrs and GoFundMe.

Some of the most successful crowdfunded projects and the money they raised

Pebble (*smartwatch*)	$20.3m
Coolest Cooler (cooler box)	$13.2m
Kingdom Death: Monster 1.5 (game)	$12.3m
The World's Best Travel Jacket (clothing)	$9.19m
Exploding Kittens (card game)	$8.78m
Sondors (electric bike)	$5.8m
Super Troopers 2 (movie)	$4.4m

Cryptocurrency

This is a form of digital currency – one that can be transferred electronically from one user to another anywhere in the world – but much more (and by that, I mean unbelievably more) complicated than **PayPal**, for example.

The fundamental difference is unlike banks and **online** financial services, there is no single governing body overseeing and verifying cryptocurrency transactions. It's administered by a far-flung **network** of volunteer programmers and computers that maintain all of the records. Now that might sound decidedly dodgy, but the system works and offers these advantages:

* no middleman (i.e. a bank) charging transaction fees;
* it can work outside traditional banking hours so transactions are virtually instant;
* no central storage for account information means all user data is safer from **cyber attack**.

Trying to explain and understand how cryptocurrencies work involves terms such as **blockchain**, peer-to-peer transaction, keypair, **network** node verification and mining – and will give both of us headaches and probably nosebleeds. To be honest, if you're reading this book, it's probably best if you ignore cryptocurrencies and just stick with your Post Office Savings Account or an ISA.

NB Cryptocurrencies are secured in a virtual wallet, although 'secured' may be a loose term since it's estimated that over $9 million per day gets scammed from cryptocurrencies.

See also: **Bitcoin**

With over 1500 cryptocurrencies available, it's difficult to discriminate between them, which is why some have used celebrity endorsements from people including Paris Hilton, Jamie Foxx and Floyd Mayweather to help them stand out. One of these endorsed cryptocurrencies was Centra. After a successful launch the founders were later arrested and charged with fraud. It was later discovered that the CEO of Centra didn't actually exist and his **profile picture** on its **website** was actually of a random Canadian physiology professor.

Cyber attack

This is an attempt to **hack** into an enemy's computer systems or entire computer **network** for malicious intent. It could be something as simple as installing **malware** on an individual's PC, to destroying a whole nation's infrastructure.

Cyber attacks during wartime are known as cyber warfare, and usually involve acts of espionage and sabotage.

See also: **cybercrime**, **cyber terrorism**

Cyberbullying

Back in the day, bullies would attack their victims physically or verbally and you'd know exactly who they were . . . mainly because the bully was the person stealing your lunch money or humiliating you in public. Cyberbullying, which is any form of bullying that takes place **online**, typically using **social media** sites, is far easier to get away with since it can be completely anonymous, causing additional anguish to the victims. What's more, it can take place twenty-four hours a day, seven days a week and go **viral** very quickly, causing yet more distress. Cyberbullying typically involves sending/**posting/sharing content** that is harmful, false, intimidating, tormenting, threatening, embarrassing or humiliating. Or all of these.

According to recent surveys, over 40 per cent of children say they have been bullied online, with one in four claiming it has happened more than once; 70 per cent said they had seen evidence of cyberbullying.

Cybercrime

Basically, this is any criminal activity that involves a computer and a **network**. While most cybercrimes are carried out to make criminals rich, some are done out of malice (e.g. spreading a **virus** or **malware**) or to facilitate other illegal activities (e.g. using the network to distribute illegal images or information).

Cybercrime is one of the fastest growing criminal activities in the world and can be carried out against individuals, for example through obtaining passwords in order to access bank accounts, or against organisations, e.g. obtaining confidential information, which the criminals can use to their advantage or sell on.

Cybercrimes against governments or nations fall into the categories of **cyber attacks** or **cyber terrorism**.
See also: **spyware**

Cyber terrorism

This is the use of the internet by politically motivated individuals or groups to either intimidate those who don't share their views or to cause disruption and damage to a country's infrastructure. Whatever the reasons for cyber terrorism, the aims are the same: to destabilise and cause fear and anxiety among the general population. Examples of cyber terrorism include **hacking** into computer **networks** in order to disrupt infrastructure such as financial institutions, telephone companies, the railways, air traffic control, etc.
See also: **cyber attack**

Cyber vetting

Back in the day, you applied for a job and, if your employers liked you, they would make you an offer and then maybe bother to check your references. Nowadays employers are far more inquisitive (cynical) about candidates' claims and many carry out what's known as cyber vetting, also known as **online** vetting. This basically entails checking you out online for two reasons.

The first is to see if you've lied on your CV. For example, did you really work for NASA or as Secretary General of the UN?

The second reason is to gain an indication of your character. By trawling through **social media posts** and photographs it's quite easy to see if you're dishonest/lazy/violent/ignorant/self-absorbed – or are a drunkard/racist/Luddite/criminal; basically, whether you have any negative personality traits.
See also: **digital footprint**

From DAB radio *to* DVR

DAB radio

You may already have one of these in your car or house; the acronym stands for digital audio broadcasting, a way of transmitting radio signals digitally. In the same way that TV in the UK switched from analogue to digital in 2012, the same thing will happen to analogue radio broadcasts (AM and FM). No timescale has been set yet, but it's likely to be in the next few years.

So why do we need to switch? Well, the main argument is that it will provide a greater choice of radio stations (digital signals take less space on the airwaves); it also means that the stations can transmit data alongside audio so you can see, for example, who's singing/playing the music you're hearing. In addition, some DAB radios allow you to pause, rewind or record radio shows.

The downsides are that DAB radios are relatively expensive, the service is patchy across the country and many stations currently broadcasting on the DAB **network** have an inferior sound quality compared to FM. It's like the old joke about going to the wedding of the chairman of a DAB radio station. The service was great but the reception was terrible ...

Dark web

To understand what the dark web is, think of the **World Wide Web** like a gigantic museum or gallery. It might take ages traipsing round, and you'll probably need to ask for directions and eventually get fed up looking, but all the stuff on the display is accessible. What you can't see, though, is an equally incredible amount of stuff hidden from the public in the basement.

Think of the dark web as the World Wide Web's basement. In this case, though, it's a depository for all things, well, dark.

And for 'dark', read 'dodgy' or 'dubious'. The dark web is the go-to place for things such as buying crystal meth, fake passports or Semtex, to hiring a hitman or learning how to cook people.

Seriously.

You can't just log on though and view the dark web. All the **content** is encrypted and before you can access information you have to know a **website**'s actual **URL** and **download** an **encryption** tool called TOR.

So, sleep safe in the knowledge that when you're innocently searching for 'Will frost affect my gardenias?' it's highly unlikely you'll accidentally stumble across a **forum** on cannibalism.

See also: **deep web**

Database

This is information held on a computer that can be accessed and organised in different ways. Databases can include information on the customers or users of a service, but can also include information on objects; for example, a shop would have a database of its customers so it can **email** them information about a sale, and it would also have a database of its stock so it can tell when items are running low. Every entry in a database is called a 'record' and every different way you want to sort these records is called a 'field'. For example, your doctor's surgery might have these fields for its patient database: name, address, telephone number, age, allergies, current treatment, date last seen, etc.

For security, databases are usually (or should be) password protected.

Deep web

Sometimes this term is used interchangeably with the term **dark web**, but strictly speaking it's not the same thing. The deep web is basically anything on the web that can't be found by a typical **search engine** such as **Google**, **Bing** or **Yahoo!** This includes nefarious or illegal **websites** on the dark web – but also a huge number of innocuous websites containing things such as **databases**, **files** stored in the **Cloud** or in a **Dropbox** account – along with business **intranets** or websites hidden behind **paywalls**. It's estimated that the deep web is 400 to 500 times larger than the regular web (sometimes called the 'surface web').

Device

See: **mobile device**

Digital assistant

These are devices that use voice control technology to provide information or carry out tasks using speech recognition and natural language. The best-known examples are **Amazon**'s **Alexa**, **Google**'s Google Assistant, **Apple**'s **Siri** or **Microsoft**'s Cortana. The built-in **software** can also interact with **smart home** devices i.e. to control lighting, change TV channels, etc.

They're also known as an intelligent personal assistant, AI assistant, digital voice assistant or virtual assistant.

Digital footprint

Your digital footprint is all the information that exists about you on the internet as a result of your **online** activity. This includes your web browsing history, all your **posts** on **social media**, **chat rooms** or **forums**, any **online** shopping orders and, of course, your **emails**. All these activities leave a permanent trail across the internet that people can access. This includes potential employers or recruitment agencies, many of whom now carry out **cyber vetting** to check your background and to verify your CV – or criminals who might try and steal your identity.

To make sure that your **profile** is seen by just the people you want, make sure you select the right privacy settings on your social media accounts (these generally allow either friends/contacts or the general public to view your profile and your posts).
See also: **cybercrime**, **doxing**

Domain name

This is the **website** address (or **web address**) for a company, an organisation or even an individual. It's what you type into your **browser** to find a website and is preceded by the letters 'www' e.g. www.facebook.com, www.youtube.com, www.bbc.co.uk.

A domain name is not the same as a **URL**, but it's close.
See also: **IP address**

You can tell what type of website you are on by looking at the last part of the domain name. The following are the main ones used in the UK but there are hundreds more:

- .com or .co.uk signifies that the website is a business (it could be anything from a one-man business to a multinational corporation)
- .org signifies a not-for-profit organisation such as a charity
- .edu signifies the website is an educational institution of some sort
- .ac.uk is used by UK universities

Domotics

Take the Latin name for home, *domus*, and combine it with the word *robotics* and what have you got? A really dreadful alternative new word for **smart homes**, or home automation, that no one really uses.

Dongle

Stop sniggering at the back.

In this sense, a dongle is a small device that plugs into a **USB** port on your computer or **laptop** to provide different types of internet connection. For example, a USB **modem** dongle provides **mobile broadband** via a 3G or **4G** mobile **network**. **Streaming sticks** are also dongles, enabling a non-**smart TV** to access the internet and streaming video services.

Why is a dongle called a dongle?

The truth is, no one knows for sure. These are three reported explanations:

1 It's a completely made-up 'nonsense' word for a piece of technology, such as gizmo, widget or doodad.

2 It's a corruption of the word 'dangle', to reflect something sticking out of a computer port.

3 A 1992 ad for the device by Rainbow Technologies in *Byte Magazine* claims it was invented by a Mr Don Gall.

Download

Used as a verb or a noun, this is a broad term that means transferring something you find on a **website** or **network** to your own computer or **mobile device**. This could be a document, a song, a **podcast** or an image, for example. The time it takes to download your **file** depends on the size of the file itself (larger files take longer) and the quality of your internet connection or **WiFi** signal. It can also be affected by how many other people are trying to access the data at the same time.

See also: **upload**

Doxing

This is stalking someone **online**, gathering private and personal information about them from various sources and then publishing this data without their prior knowledge – or indeed, consent. Doxing can be carried out by individuals or companies (typically the media) for a variety of purposes, e.g. revealing the identity of someone making anonymous (and controversial) **posts** or revealing the identities of people involved in criminal activities.

The practice isn't illegal per se as long as the information is available in the public domain – usually drawn from **social media** or publicly available **databases**. Of course, if the information is used for things such as blackmail or harassment, then that's a whole other issue. The best way to avoid being doxed is to reduce your **digital footprint**.

Drone

Also known as quadcopters, drones are unmanned miniature helicopters that use four rotors to provide lift and control movement. Technology has fuelled their popularity; controlled by **smartphones**, they use **GPS** technology for guidance and most have built-in cameras, some with **4K** definition and gimbals that allow the image to remain rock steady while the drone moves about. Although once exclusively used by movie makers or by industry and the military for reconnaissance and surveillance, drones nowadays are increasingly used by the media to get aerial paparazzi shots – and wedding photographers who try and turn what seems like the most uninspiring of nuptials into what looks like a Hollywood blockbuster.

Controlling a drone is harder than it looks – which is why some more advanced models have both an 'obstacle avoidance mode' and a 'return to home' mode – essential when it disappears from view.

Amazon is exploring the viability of using drones to deliver packages weighing under 5 lb – which accounts for 90 per cent of its sales.

Dropbox

Despite the name, this was not a furtive method for exchanging state secrets during the Cold War; instead it's a personal **Cloud** storage service, mainly used for sharing **files** with other people. That said, although it doesn't have the romanticism associated with the world of espionage, it is very clever. How it works is simple; you create a Dropbox account then **upload** the files you want to make available to others. To **share** a file, you simply add the recipient's **email** address and a **hyperlink** is automatically sent to them, giving them access. Dropbox is available for personal or business use; it's free with limited storage – and you can buy additional storage for a monthly subscription. As for security, well, files are protected with 256-bit AES **encryption** (and that's good).

NB If you don't want to, or don't need to, share files then Dropbox is also a great **back-up** service. Once you've uploaded your files you can access them from anywhere on any device (providing you've got an internet connection).

Although Dropbox is a US company with over 500 million users worldwide, the name has become generic for file-sharing services such as OneDrive from **Microsoft** and **Google** Drive from, well, you know.

DVR

A digital video recorder allows you to pause or rewind live TV viewing and record TV broadcasts on to a **hard drive** for later viewing. It also gives you the opportunity to easily record whole seasons of a TV show.

See also: **TiVo**

From Easter egg *to* Extranet

Easter egg

Not the seasonal confectionery that's 35 per cent chocolate and 65 per cent packaging, these Easter eggs are supposedly more satisfying, being intentional in-jokes, secret features, messages or images hidden in computer games or films by their programmers/directors. Their name comes from the fact that you have to search for them like a traditional Easter egg hunt.

In the case of computer games, to access Easter eggs you usually have to follow a series of convoluted instructions (you'll find these online) such as pausing on certain scenes, pressing your left or right keyboard arrows, clicking on a specific area, unpausing the disc, pausing it again after a certain number of seconds, typing in a secret word, then repeating the process at another part of the disc. This laborious process could result in the reveal of a secret game (which is quite interesting) or the ability to change the background colour of the **menu** (which, it has to be said, is not).

In the case of films, Easter eggs aren't hidden as such; they're on screen to be spotted by eagle-eyed viewers. For example, in *Raiders of the Lost Ark*, hidden among a set of ancient hieroglyphics are depictions of *Star Wars* droids C-3PO and R2-D2; in the film *Fight Club*, there's a Starbucks coffee cup in every scene.

You can find a comprehensive list of DVD Easter eggs at: www.hiddendvdeastereggs.com

eBay

The world's biggest **online** marketplace where you can buy and sell almost anything either via an online auction or at a fixed price (what eBay calls 'Buy It Now!'). The service is free to use for buyers, but sellers are charged fees for listing their items. But you already knew all that, didn't you?

What you might not know, however, was that the site's original name was AuctionWeb. The founder, Pierre Omidyar, wanted to change the name to Echo Bay after the recreational area near Lake Mead in Nevada; however, the **domain name** echobay.com was already registered to a Canadian mining company Echo Bay Mines. Pierre simply dropped the 'cho', and ebay.com was born.

Ten other things you may not know about eBay

1 It's estimated that eBay has 170 million active users.

2 The first item sold was a broken laser pointer.

3 In 2004, a partially eaten, ten-year-old grilled cheese sandwich that bore the image of the Virgin Mary sold for $28,000.

4 In January 2006, Briton Leigh Knight sold an unwanted Brussels sprout left over from his Christmas dinner for £1,550 in aid of cancer research.

5 In May 2006, someone from Brisbane, Australia, attempted to sell New Zealand at a starting price of A$0.01. The price had risen to $3,000 before eBay closed the auction.

6 In June 2002, Disney sold a used Monorail for $20,000.
7 In December 2004, water that was said to have been left in a cup Elvis Presley once drank from at a concert in North Carolina in 1977 was sold for $455.
8 After a failed marriage, in June 2008 Englishman Ian Usher from Durham auctioned his 'entire life' including his house, all his belongings, an introduction to his friends and a trial at his job. When bidding closed, his 'life' had sold for $384,000.
9 The most expensive item sold to date was a 405-foot superyacht bought by Russian billionaire Roman Abramovich in 2006 for $168 million.
10 Items that cannot be sold on eBay include partially used cosmetics, used underwear, radioactive waste, personal information, human remains and body parts (excluding clean skeletons and skulls).

E-book

A digital version of a printed book that can be **downloaded** and read on a computer or **tablet** via free e-book reader **software** or on a dedicated **e-reader**.

See also: **e-ink**

After a rapid rise in popularity fuelled by the launch of **Amazon**'s best-selling **Kindle** e-reader in 2007, e-books are now in decline, while sales of printed books are in the ascendancy once more.

E-cigarette

A device that's supposed to wean smokers off conventional cigarettes, or at least provide a slightly healthier alternative, the electronic cigarette is designed to allow smokers to inhale nicotine without inhaling two of the main toxins of the smoke: tar and carbon monoxide.

Looking like cigarettes, they work by heating and creating a vapour from a solution that contains nicotine, some other chemicals and flavouring. Since there's no burning involved, there's no harmful smoke.
See also: **vaping**

E-ink

A type of electronic display characterised by high contrast and clarity, and a wide viewing angle. Most commonly used for **e-books** to recreate, as near as possible, the experience of reading a real printed book.
See also: **e-reader**

Email

You're probably thinking, 'Look, I know what email is, duh! Don't treat me like a fool.'

I'm not, but there might be some readers who don't know what it is, so this entry is for them, OK? For the rest of you, skip this and go look up something else you find more challenging, maybe **'jailbreak'**, **'net rep'** or **'viral marketing'**.

So . . . for anyone who doesn't know what email is, it's one of the most prevalent forms of communication today, replacing written correspondence and also telephone conversations. Think of it as sending and receiving letters over the internet, but much, much, much better. To use email you need a computer/**mobile device**, an internet connection and an email account (which will be provided by your **ISP** or **webmail** provider).

Your email account will give you a unique email address that consists of a name (usually your actual name but it could be a nickname or something random), followed by the @ symbol and then the name of the email service provider, e.g. Barneybear579@teknetticom.co.uk or Misspolly@talkasaurus.com.

Before you can do anything with email you have to

first open your email **program**. Once this is done:

To send an email: enter the email address of who you want to write to, enter a subject name for your message (e.g. 'How are you?', 'Send money', 'You're having my baby'), then compose your message and press 'Send'.

To read an email: click or double click on the message and it will open.

To reply to an email: once the message is open (see above), click on 'Reply', compose your response and then press 'Send'.

To receive emails: click on 'Receive messages' or 'Get mail' – or similar.

It's e@sy (see what I did there?).

See also: **attachment**, **Bcc**, **Cc**, **spam**

Advantages of email over letter writing

* Faster (communicate to anyone anywhere in the world virtually instantly)
* Convenient (send and receive messages any time you're connected to the internet)
* Flexible (send the same message to as many people as you want simultaneously)
* Inclusions (you can attach things such as photographs, documents, songs or videos to your message)
* Permanence (all your emails sent or received are stored on your computer or in the **Cloud** so you can easily refer back to them)
* Unlimited (send as many emails as you want; the only cost is your monthly internet subscription, which you need anyway to access the **World Wide Web**)
* You don't need to find that book of stamps you're sure you had in your purse, on your bedside table or in that drawer with the pens, old batteries, the tape measure and scissors.

Emoji

Unlike **emoticons**, these are actual images you can insert in **online** messages to add emphasis. There are vast libraries of emojis in most messaging **apps**, and they can also be accessed from the keypad of your **mobile device**. They include facial expressions, common objects, animals, places, vehicles, flags, types of weather, foods – anything really, including a smiling alien and someone painting their nails (difficult to know where you'd use both of these in the same message).

Like emoticons, the unwritten rule of using emojis is don't overuse them and don't use them inappropriately. For example, if you're **texting** a friend with: 'Meet you at the restaurant at 7 p.m.', there's a high likelihood that the recipient will know what a restaurant is; there's really no need to add eight different food emojis.

Equally, if you're relaying the fact that someone died after a long illness it's probably not in the best taste if you add the ghost emoji.

Emoticon

Largely superseded by **emojis**, an emoticon is a depiction of a face made from the letters, numbers and symbols on your keyboard or keypad. They're used to represent your mood or intended tone in a message. In addition, the use of a strategically placed emoticon in a **text** or **email** can allow you to speak the truth without appearing too impolite, e.g. *Great to see you. It would have been even better if you hadn't kept me waiting in the rain for five hours* ;-)

Five common emoticons

:-)	I'm happy
:-(	I'm sad
;-)	I'm winking

:O	I'm surprised
:-###..	I feel sick/I've been sick

Emoticons have been called 'the smallpox of the internet'. If you're going to use emoticons, use them sparingly. The generally accepted limit is two per text or email.

Encryption

This is the process of scrambling computer data so it can be sent securely, i.e. no one apart from the recipient will be able to read it. It's done by encryption **algorithms**, which turn the information into a complicated code that can only be unlocked by the authorised recipient using a special decryption 'key' (think of the process like translating the information into a special language that only you and the recipient can understand).

There are different types of encryption with different levels of effectiveness. These are measured in 'bits'. The higher the number of bits, the harder it will be for a **hacker** to crack the code. For example, a 5-bit key has 32 possible combinations (i.e. not very secure), whereas a 256-bit key has 115,792,089,237,316,195,423,570,985,008,687,907,853, 269,984,665,640,564,039,457,584,007,913,129,639,936 possible combinations (i.e. very, very, very, very, very, very secure).

If you visit a **website** with a **URL** than begins 'https://' then that indicates any data transferred will be securely encrypted. You should find it on sites that take online payments.

EPG

Normally accessed by the button 'EPG' or 'Guide' on your remote control, an electronic programme guide is an on-screen listing of all of the programmes available on your TV. It's also the **interface** by which you select a channel or record shows via a **DVR**. To save time searching for the shows you like best, you can usually view a list of programmes by genre, e.g. movies, news, documentaries, etc.

E-reader

Also called an **e-book** reader, this is a handheld device similar to a **tablet** but designed primarily for reading electronic books (e-books) and magazines – as well as providing the ability to purchase and **download** titles. Popular e-readers are **Amazon**'s **Kindle** and models from Kobo and Nook.

E-reader sales are in decline due to a lack of technical innovation since they were first introduced in 2007, the fact that more people are reading e-books on larger **smartphones** and tablets – and a renaissance in sales of printed books.

Ethernet cable

Looking like a phone cable on steroids, this is a type of cable typically used to connect computer **networks** for businesses. Domestically it's more usually seen connecting a computer or **smart TV** to a **broadband router**.
See also: **Cat 5, 6, 7**

Etsy

A sort of **online** craft fair; a **website** where you can buy and sell handmade or genuine **vintage** items including art, clothing, accessories, shoes, jewellery, beauty products, memorabilia and toys – along with a whole assortment of craft supplies. What makes Etsy different from other online retailers is that the majority of its items are handmade and therefore unique, and most sellers are independent (i.e. they actually made the items they are selling). Like **eBay**, Etsy charges a listing fee for each item offered for sale and takes a commission on every sale.
Visit: www.etsy.com

Extranet

This is an **intranet** that is partially accessible to people from outside the company or organisation; built-in security features provide different degrees of access depending on whether you're an employee or an external party. Examples of extranets include a construction company sharing information on a project with all its various sub-contractors, or a school providing access to parents so they can get school news or view their child's exam results or reports.

From Facebook *to* Freemium

Facebook

This is the go-to place to let your friends know that you're leading a fantastically fulfilling life that consists of great parties, exotic holidays, attractive friends, successful children and talented pets – and then finding out that they are doing exactly the same (but more so).

If you don't already know what Facebook is (even though you're reading this book the chances are that you're one of its 2.2 billion worldwide monthly users), you've probably heard of it.

Launched in 2004, Facebook was one of the first **social media/social networking** sites. Today it's the most widely used; 1.45 billion people on average access Facebook each day to **post** comments, **share** photos, videos and **hyperlinks** to news features or advertisements, play games and even **stream** live video. You can adjust privacy settings that determine whether your posts and personal information about yourself are shared among your select group of friends and family or publicly. There are also Facebook groups you can join to share information about a topic you have in common (these work just like an internet **forum**).

Facebook's not just for 'ordinary' people wanting to keep in touch with/impress their friends and family. If you're a company, an organisation, a celebrity or just someone with a big ego, you can set up what's called a Facebook page to promote yourself, your business activity or your brand. Unlike personal **profiles** where you gain 'friends', these pages gain 'fans'.

See also: **block**, **unfriend**, **Zuckerberg**

Twelve things you may not know about Facebook

1 The maximum number of friends you can have is 5000.

2 It was originally limited to members of Harvard University, where the site was created.

3 In 2016 it was estimated that 600,000 hacking attempts were made on Facebook accounts daily.
4 It's available in 101 different languages.
5 It's estimated that there are more than 30 million dead people still on Facebook.
6 . . . And 83 million fake profiles.
7 The first major investor to back Facebook was Peter Thiel, co-founder of **PayPal**. He invested $500,000 into the young company in 2004 and later sold his stake in the company for more than $1 billion.
8 Eight new users join every second.
9 The most popular time people use Facebook is midweek between 1 p.m. and 3 p.m., and the average time spent per Facebook visit is twenty minutes.
10 The 'like' button was originally going to be called the 'Awesome' button.
11 The most shared **content** is videos.
12 According to a study, one in three people feels more dissatisfied with their lives after visiting Facebook.

Faceprint

This is a digital representation of your face, used in conjunction with **biometric software** for identification/authentication. A faceprint is as individual as a fingerprint but **facial recognition** isn't 100 per cent foolproof. **Apple** claims that with its Face ID software there's a one-in-a-million chance of someone else's face being able to unlock your phone.

FaceTime

A way to make free audio and video calls to other **Apple** users. An alternative to **Skype** or **Google Duo**, this system is integrated into all Apple devices with forward-facing cameras, so you don't need to set anything up. Just sign in with your Apple ID and you're all set to go.

Facial recognition

Also known as Face ID (the name of **Apple**'s own system), this is a method of recognising an individual by comparing their live image with a stored **biometric** record of that person, called a **faceprint**. The system is used for identification whether this is you being able to make an electronic payment or for security organisations to find a 'person of interest'.

It works by **software** analysing your image and measuring eighty 'nodal points' on your face. For example, the distance between your eyes, the width of your nose, the shape of your cheekbones. This data is then stored as an accurate record of your face. When you want to unlock your phone, for example, you look at it, an image of your face is compared to this stored data and if the nodal points match – you're in!

Fake news

Ever since man witnessed stuff going on, there's been the accurate reporting of events (the news) and inaccurate reporting of the same thing (rumours). In recent times, however, the line between the two has become blurred, such is the speed that news can be made and **shared** via **social media**. Nowadays, rumours – some without any foundation whatsoever – can become established and ingrained as news; this is what critics of those reports call fake news. It's a term that's thrown around a lot and it doesn't help to distinguish what's just a little bit fake (e.g. by exaggeration or for satire), slightly fake (maybe a few facts have been omitted) or absolutely, positively fake (no sources, no foundation in reality and what might be considered the ravings of a madman).

Fake news is a concept that rose to prominence with the 2016 US election with both the Republican and Democratic parties responsible for spreading news stories to discredit their rivals. Analysis by **BuzzFeed** found that fake news stories were shared more, and drew more engagement, during the final three months of the election campaign than reports from reputable media such as the *New York Times*, the *Washington Post* and CNN.

Such is the extent and impact of fake news that in December 2016 **Facebook**, working with users and third-party fact-checkers, decided to flag up fake news stories that were **posted** by putting a red warning icon next to them. It dropped this policy within a year after research found that putting a warning sign next to a news article may actually embed deeply held beliefs even further – the complete opposite to what was intended.

Fanboy/fangirl

A slightly derogatory name for someone who's more than just a fan; someone who's obsessively and hopelessly devoted in a very over-excited way to something. That something is usually a singer or band, a product or manufacturer (e.g. **Apple**), a video game, a TV or film franchise that will typically involve science fiction, boy

wizards, vampires or inner earth). Regardless of any deficiencies or shortcomings in the subject of their affections, fanboys and fangirls tend to treat all critics as **haters**.
See also: **cosplay**, **fanfic**

Fanfic

For some obsessive fans, it's not enough for them to merely read books or watch TV series/films about their favourite characters; they feel compelled to write original stories or scripts that use the same characters and settings. This fan fiction (shortened to fanfic) is unauthorised and is typically created for fellow fans to read and appreciate, being circulated on **forums** or dedicated fan-fiction **websites**. Reaction to fanfic by the creators or copyright holders of the original work can range from appreciation to indifference to legal action.
See also: **fanboy/fangirl**

File

Computer files are like the traditional files you'll find kept on a shelf or in a filing cabinet. Like their paper counterparts, they can contain any sort of information (or data). For example, a computer file might be a document, a photo, a **PDF**, music, a video – or a **software program**.

A computer file name will be followed by a full stop and then three or four letters/numbers. This is known as the file extension and it signifies what kind of file it is.

File extension	Type of file
.doc	**Microsoft Word** document
.jpg or .jpeg	Photograph
.mov	Video file
.mp4 or .mpeg4	Video file
.mp3	Audio file
.pdf	PDF

.ppt	Microsoft PowerPoint document
.wav	**Windows** audio file
.xls	Microsoft Excel file

Filter

Nothing to do with **e-cigarettes**, these are digital effect filters that change the look of photographs or videos **posted** on a host of **social media platforms**. Although the number of filters (and their corresponding names) vary greatly on these platforms, the sort of effects that can be achieved include deepening shadows, brightening highlights, making photos colder or warmer, giving them a vintage feel, intensifying some colours while leaving others untouched, saturating or desaturating the whole shot, softening skin tones and smoothing wrinkles (essential for portraits) . . . you get the idea. **Instagram** and **Snapchat** also offer a host of face filters that let you add things such as a bunny nose and ears, nerd glasses, costumed ball masks, crowns and tiaras to your **selfies**.

It's hard to think of a more First World Problem than 'I can't seem to add the jewelled face mask to my selfie.'

Fire TV Stick

Amazon's own media **streaming** device; the latest version features **Alexa** voice-activated remote control.
See also: **streaming stick**, **video on demand**

Firewall

Think of this as a barrier between the internet and your computer, but rather than a physical barrier it's usually a piece of **software** that provides security in two ways; it stops unauthorised users accessing your data and prevents undesirable data, e.g. **malware** and **viruses**, from getting through.

Some operating systems have firewalls built in and

companies sometimes have a firewall system in place but that doesn't mean they are foolproof. Malware can still get through via opening **email attachments** or email **hyperlinks**.
See also: **hacking**

First person shooter

A video-game genre where you experience action from the point of view of the character (and by 'action', read 'violently killing people or creatures'). First person shooter (FPS) games tend to be fast-paced, ferocious, violent and bloody.

Pacman and Crash Bandicoot were never like this.

Fitness band

Also called fitness trackers or activity bands, this is a type of wearable computer like a **smartwatch**, but one which is designed to mainly monitor your fitness. They usually have a much smaller display than a smartwatch (or no display at

all) and primarily provide readings of things such as calories burned, the distance walked or run, general activity levels, heart rate and, in some cases, even the quality of your sleep. They connect wirelessly to your **mobile device** or computer and synchronise information/results so you can check your fitness progress (or lack of it). As their name suggests, many are lightweight bands worn on the wrist or ankle while some just attach to items of clothing.

Flaming

Say you're watching a Coldplay video on **YouTube** and you **post** a comment about how the band's not as good as Snow Patrol. To your surprise it's met with a barrage of insults from Coldplay **fanboys** who vehemently disagree (and helpfully point out the band's seven Grammy wins). This is not good-natured banter about meaningful lyrics or musical style, but messages that are as hostile as they are sweary. This inflammatory interaction is called, logically, flaming.

Flaming can take place on any **social media platform** where there's an opportunity for comments to be swapped (e.g. **forums**, **Twitter**, **newsgroups**). It can be the exchange of insults back and forth with someone, or multiple people 'ganging up' on one individual. A continued dialogue with insults being thrown around is called a flame war.

Flaming usually happens when the discussion is about hot topics such as politics, religion, immigration, feminism, climate change, ethical issues – or differences in musical tastes. Flamers, however, can be as irrational as they are prejudiced, with the result that no topic is flame proof.

See also: **hater**, **troll**

Flash drive

This is a small, low-cost storage device, about the size of your little finger (if you've beaten your little finger flat with a hammer). Also called a **USB stick**/drive or a memory stick, it's taken the place of the 3.5-inch floppy disc as one of the most popular ways to physically transfer **files** from computer to computer. Nowadays the smallest flash drives available tend to have a capacity of 8 GB . . . the equivalent of over 5000 old-school floppy discs.

Flashmob

You may well have experienced this. You've been on the concourse of a busy station minding your business, waiting for the delayed 19:50 to Basingstoke when a crowd suddenly appears around you and then, without warning, launches into an impromptu a capella rendition of 'Don't Stop Believing', 'Let It Go' or 'Single Ladies', before dispersing as quickly as they arrived.

This was a flashmob, a group of (usually) unconnected people who assemble apparently spontaneously and randomly in a public place to perform a song or dance and then leave.

Flashmobs tend to get their instructions via **social media**. As for why do they it, well, for many it's for entertainment or artistic expression. For others, it's just a case of, 'Why not?'

Flickr

A photo-sharing and **hosting** service where users can **upload** their own photos and videos for others to see and enjoy (and perhaps purchase). Think of it as **Instagram** for professional photographers or at least serious amateurs.

Approximately 1.4 million photos are uploaded to Flickr each day. The most popular camera used is the **iPhone**.

Follower

Mention 'follower' and the term can conjure up images of evil cults and fanatical disciples; in the context of **social media**, though, it's slightly less worrying. If you're interested in seeing what someone else is saying on **Twitter** then you need to subscribe to their Twitter account (also known as their 'Twitter feed'). Once you've done this you're known as one of their followers, which means you can see that person's **tweets** and they can also send you direct messages.

You can be a follower of an individual, a company or an organisation – and it's also possible to buy Twitter followers in order to boost your apparent popularity.

Celebrities with the most followers (at the time of publication)

Katy Perry	107 million
Justin Bieber	104 million
Barack Obama	102 million
Rihanna	88 million
Taylor Swift	83 million
Lady Gaga	77 million
Ellen DeGeneres	76 million
Cristiano Ronaldo	75 million
Justin Timberlake	64 million
Kim **Kardashian**	59 million

Forum

Forums are like **chat rooms** but with two main differences. The first is that the participants don't have to be **online** in real-time to start or join discussions; they can **post** a message on any topic at any time on any day. The second difference is that forums tend to be quite specialised. You don't join one if you want to meet people or have a general chitchat. You do join one if you want to

find out if anyone knows how to adjust the valve clearances on a 1995 Ford Mondeo.
See also: **moderator**, **newsgroup**

4G

Not to be confused with **4K**, 4G is the fourth generation of the mobile phone network; an industry standard that allows wireless internet access at a much faster speed.

4K

A term used to describe an ultra-high definition (Ultra HD) TV screen, so-called because it has 4000 **pixels** horizontally. You can tell if you're watching a 4K film or broadcast by the fact that it massively accentuates every actor's blemishes to the point of distraction – so much so that it's a struggle trying to appreciate a sensitive, heart-wrenching monologue when you find yourself captivated by the leading actor's blackheads.

4K has four times the picture resolution of previous 1080p screens, which were merely 'high definition', but before you dash out and buy a state-of-the-art 4K TV to enhance your viewing pleasure, it's worth noting that 8K screens are just around the corner.

Freemium

A way of pricing services, especially from **online** companies, where basic services are provided free of charge, while more comprehensive or advanced features have to be paid for with a subscription. An example is **Spotify**, which is free, but you have to endure commercials between songs. If you pay a monthly subscription you have more tracks available to you, better quality sound – and no commercials.

From Generation X *to* Grindr

Generation X

The generation before Generation Y (also known as **Millennials**), the term Generation X (or Gen X) is used to describe those born between approximately 1965 and 1980. They have been characterised as the '*Friends* generation' (as in the TV series): unfocused, self-involved, aimless but fun. Generation X is sometimes also referred to as the 'MTV Generation'.

Generation Y

See: **Millennials**

Generation Z

Also called Gen Z, Post-millennials or iGen, this term is used to describe the generation after **Millennials**, but there's no consensus among sociologists or researchers about what birth years it actually refers to . . . which makes the whole term a bit meaningless. The approximate birth years for Generation Z are the mid-1990s to the mid-2000s. Give or take.

GIF

See: **animated GIF**

Google

Do I really need to explain what Google is?

Really?

REALLY?

Well, OK then. Founded in 1998, it's the world's most popular **search engine**, being used for almost 91 per cent of all web searches. It's estimated that around the world, 40,000 Google searches are performed every second; that translates to over 3.5 billion searches per day.

Ten things you might not know about Google

1 The original name for Google was BackRub.
2 It's been named as the company that most often changes its logo.
3 Devotees of Google can worship the company at the Church of Google website.
4 When you search for 'Askew', the results slant to the right.
5 Google rents 200 goats (and a goat herder) to 'mow' the lawn and keep the weeds around its headquarters under control.
6 The first example of Google being used as a verb ('to google something') on TV took place during an episode of *Buffy the Vampire Slayer* in 2002. It wasn't until 2006 that it was added to the *Oxford English Dictionary*.
7 Dogs are encouraged in Google offices as long as they are of a friendly disposition and have strong bladders. Cats are discouraged due to the number of dogs present.
8 The company's unofficial motto is 'Don't be evil'.
9 Google's first storage was ten **hard drives** in a Lego casing. The Lego allowed for easy access and expansion of the storage.
10 To protect itself from possible negative publicity, Google owns the **domain** googlesucks.com.

Google Duo

Google's video chat **app**; its version of **Skype** or **FaceTime**, and available to use on both **Android** and **iOS** devices.

Google Play

The one-stop shop for **apps**, games, music, movies, **e-books**, newspapers and magazines, designed to run on Android devices. The whole Google Play store can be accessed via its website or via the Play Store app.
Visit: https://play.google.com/store

Google+ (or Google Plus)

Put simply, Google+ (or Google Plus) is a version of **Facebook**. As **social media** go, **Google** was a little late to the party, launching in June 2011 (Facebook's been around since 2004). The principles of both sites are the same, although Google+ has some unique features such as Hangout, which provides the ability to have up to ten people join you for a video chat (useful for businesses). Google is cagey about how many active users Google+ has, but it's struggled to compete with Facebook. Some reports claim it has 343 million active users (versus Facebook's 2.2 billion monthly active users); however, other studies have indicated that just 9 per cent of Google's users actually **post content** on the site. It's visited by just 22 per cent of all **online** adults per month.

Google Translate

This is a free-to-use translation service developed by **Google** that can be accessed via its **website** and also as an **app** for **Android** and **iOS** devices. It supports over a hundred languages from Albanian to Zulu and you can translate individual words, phrases and whole sentences to and from any of them. You just type in the text you want translating, select the language you want it in and then

click on 'Translate'. The translation appears instantly alongside the original text. It's so simple! (Or as they say in Swedish, *'Det är så enkelt!'*)
Visit: https://translate.google.co.uk

An **augmented reality** feature of Google Translate allows you to point your **mobile device**'s camera at any text in a foreign language and view a 'live' translation superimposed on to the image.

GPS

The Global Positioning System is a satellite-based navigation system made up of thirty-one satellites that were originally launched by the US Department of Defense but which became available for civilian use in the 1980s (technically it's owned by the US government). All you really need to know is that these satellites make **satnavs** work. The accuracy is usually to within 5 to 10 metres

depending on the device you're using but this can be reduced if the signal is blocked or reflected, say by buildings or mountains, or affected by atmospheric conditions.

Grindr

An **online dating app** aimed at gay and bisexual men and transgender people that works in a similar way to **Tinder** but has a reputation for being more often used for casual sex. It's the largest and most popular gay dating app in the world.

From Hacker *to* Hyperlink

Hacker

Any person engaged in the act of **hacking**. The mere mention of the word conjures up images of greasy-haired, unkempt computer geniuses with the nicknames 'Nightcrawler' or 'The Phantom' pounding away on keyboards in their bedrooms with hateful intent, hellbent on extorting millions from companies or accessing the nuclear launch codes from the government.

The thing is, hacking isn't black and white ... well it is, sort of. There are three recognised types of hacker:

* White Hat hackers: these are 'ethical hackers' working to test security systems and protect organisations and systems.
* Black Hat hackers: the criminals and evil geniuses.
* Grey Hat hackers: morally confused; do a bit of both ...

Hacking

Similar to 'attacking', the word hacking even sounds malicious – and in many instances it is – but what it simply means is using computer skills to gain unauthorised access to a computer **network** of a company or official organisation. This is achieved by getting around the usual security systems such as **firewalls** and passwords and

accessing the computer or network through an 'unofficial' route. It's like getting into a house without using the front-door key.

Hacking is usually associated with **cybercrime** or **cyber terrorism**; however, it can also be done to test organisations' security systems or sometimes to cause trouble and mischief just for the hell of it.
See also: **hacker**, **phishing**

Hard drive

Also called a hard disc or hard disc drive (HDD), this is the most delicate part of your desktop or **laptop** computer and, paradoxically, the most vital; it's where all your **software programs** and data are stored. Think of it like a special type of CD where data is constantly read, written and overwritten – and accessed by a small, moving arm with what's known as the read/write head. Hard drives come in various capacities, usually from 250 GB to 1 TB (1000 GB). With these rapidly moving and delicately balanced parts, knocking or dropping a computer can lead to problems accessing your data. And by 'problems', I mean catastrophic consequences.
See also: **SSD**

As a guide, a 250 GB hard drive can contain 51,000 **JPEG** photos or 64,000 **MP3** songs. A 1 TB hard drive can contain 200,000 JPEG photos or 250,000 MP3 songs.

Hardware

The name given to the physical elements of computing. This could be the computer itself or a device used in or with it. For example, a **hard drive**, **webcam**, monitor, keyboard, mouse, printer or scanner are all types of hardware.
See also: **peripheral**, **software**

Hashtag

#brexit, #metoo, #celebritybigbrother, #markleighbestauthor . . . Most commonly used on **Twitter** and **Instagram** but also used across other **social media**, hashtags are key words or phrases preceded by the hash (#) sign, and used to describe a theme or a topic. Clicking on that hashtag takes you to a page that contains all the **posts** on the same topic – making it easy to find and follow messages (or images) on that particular subject.

Anyone can create a hashtag to encourage further discussions/posts on a topic. They can be serious or silly – just remember to always precede the word or phrase by the # symbol and don't use any spaces or punctuation between words.

Hater

Someone who doesn't just dislike someone or something, but who is overly negative or critical sometimes to the point of psychotic loathing. The thing about haters is that their feelings tend not to be based on jealousy; it's more

based on wanting to undermine someone else's happiness, success or satisfaction – in effect, a desire to take them down a peg.

The term is usually used by teenagers on **social media**; haters tend to be critical of what they see as extremely overrated TV shows or bands – often for no apparent reason. The reaction from defenders of these shows/bands is to say, 'I'm going to ignore the cruel, horrible and uncalled-for comments of my critics.' This attitude is often summed up by a philosophical three-word response: 'Haters gonna hate.'

See also: **fanboy/fangirl**, **flaming**, **troll**

HDMI

High Definition Multimedia Interface: an industry standard connector that transfers high definition audio and video along a single cable. A replacement for the **SCART lead**, HDMI cables are commonly used to connect your TV or computer to devices including a **TiVo** or **set-top box**, a DVD/**Blu-ray** player, a monitor or games console.

Hipster

By the time this book comes out hipsters might have gone the way of the dinosaurs but, acting on the assumption that they will still be a 'thing', this is how to recognise them: hipsters are a sub-culture of men and women typically in their twenties and thirties who view themselves as independent and detached from mainstream consumerism, preferring 'edgy' and 'alternative' when it comes to fashion, food, music, technology, entertainment, media consumption – in fact, most things.

A more cynical description is that hipsters are those annoying, pretentious people who try a little too hard to be different.

Twelve ways to recognise a hipster

1 When it comes to fashion, their key concern is not 'Is it comfortable?', but 'Is the brand I'm wearing obscure enough?'

2 Clothing tends to be from **vintage** clothing stores and mismatched (nothing says 'I can't be bothered' more than spending two hours pre-planning your look).

3 The men look like they participate in Movember all year long.

4 Both sexes sport at least two of the following: plugs, nose studs, eyebrow rings, piercings or tattoos.

5 They **Instagram** virtually every single morsel they eat.

6 You won't have heard of any of the indie bands they like, such as Blue Gentle Dancing, Disfiguring the Kitchen and Platonic Concrete.

7 Their cameras use 35 mm film (for 'authenticity').

8 The name of each meal must be preceded by the word **artisanal** and it usually contains the following: kale, avocado, goji berries, pumpkin seeds, sweet potatoes, quinoa, more kale and an unidentified green juice.

9 Hangouts include health food stores, independent coffee shops, cycle cafés, vinyl record shops, cinemas playing obscure 16 mm films with subtitles, **retro** clothing stores and food trucks.
10 They usually have jobs with the word 'digital' or **'content'** in the title.
11 They carry a messenger bag that typically holds a half-written screenplay/song lyrics, a dog-eared copy of an inaccessible novel about existentialism and a £2500 MacBook.
12 They deny being a hipster.

Home hub

Also known as smart home hubs or home automation hubs, a home hub is the brains of a **smart home**; it controls and coordinates the communication between devices such as switches, lighting controllers, thermostats, door locks, automated blinds, wireless cameras, motion sensors, etc.
See also: **domotics**, **Internet of Things**

Homepage

Although this is one of the most basic aspects of the internet, it's made slightly more complicated because there are two different meanings.

Website homepage: if you think of a website like a book then its homepage is the contents list. It's usually the first page you **hyperlink** to and it gives an introduction to the site along with an indication of what information it contains (e.g. about us, FAQs, contact details, key personnel, main services, etc.).

Browser homepage: this is the first web page you see when you open your browser. Most people select a **search engine** as this first page as it's a convenient launch pad for accessing the **World Wide Web**. Others might choose their **email** provider, a news service, **Facebook** or the **YouTube** channel of their favourite **vlogger**. Your homepage can be any website you want.

Hosting

To give it its proper name, web hosting is the service that enables a **website** to be viewed by anyone using the internet. A web hosting company (usually your **ISP**) stores the website on its **server** so that it can be accessed by other computers. If you have a website, you pay to store your website (well, in reality, all the **files** that make up the website) on this server on a monthly or yearly basis.

Hygge

Pronounced 'hoo-ga', this is a Danish and Norwegian word for a concept that's, well, hard to define really. It's used to describe a special moment that's cosy, familiar, reassuring or charming. It's something you feel or recognise – and then appreciate. For example, lighting scented candles when you take a bath, arranging cut flowers, reading a poem – even lying in bed with freshly ironed sheets – are hygge. Apart from giving you a feeling of wellbeing, these small acts are supposed to induce a feeling of calm and reduce stress.

Just because it's an abstract notion, however, doesn't stop it being overused by people and companies trying to cash in on the concept and flog you a hygge lifestyle, i.e. anything from rugs, robes and towels to candles, soaps and mugs . . . Oh yes, and books (over 1000 with hygge in the title on amazon.co.uk).

See also: **wellness**

Hyperlink

There's no hype about a hyperlink; it's simply just a link on a **website** that lets you navigate instantly from page to page at hyper speed. A hyperlink usually appears as an underlined or differently coloured word or phrase, or even an image. Moving your cursor over it will change the arrow to a pointing hand. Click and you'll be taken to a new section within the website you're on, a different but related website, or to a **file** such as a **PDF** or an image. If this book was on a website rather than olde worlde paper, then the word PDF in the previous sentence would be a hyperlink; clicking on it, for example, would take you instantly to the definition of PDF.

Hyperlinks can also appear in **emails**.

From Instagram *to* iTunes

Instagram

A **social media app** primarily for sharing photos and videos taken directly on a **smartphone** and used by individuals and corporations/organisations alike. Just like other social media, you can interact with other users on Instagram by **following** them, being followed by them, commenting, sharing, tagging and private messaging – and 'liking' their photos. And like **Twitter**, you can add a **hashtag** to your photo so that other Instagram users can see it if it's a hashtag they're following.

The downside to Instagram is that it's quite difficult – though not impossible – to save other users' photos (the easiest way is to take a **screengrab**). Instagram is famous for its range of creative **filters** to enhance or alter images. At the time of writing there are thirty-nine different effects, but filters are added and retired regularly depending on how popular they are.

Five things you may not know about Instagram

1 It has 400 million active daily users; around 95 million photos are **uploaded** every day.

2 Its most popular photo to date was a shot of her new-born daughter **posted** by Kylie Jenner in February 2018. It has received almost 18 million 'likes'.

3 The brand with the most Instagram followers is *National Geographic* magazine – with over 91 million (Nike is second with 81 million).

4 Actress and singer Selena Gomez is the most followed individual, with almost 143 million followers.

5 The most Instagrammed food is pizza, followed by sushi.

Instant messaging

Sometimes shortened to IM, as its name suggests, this is an **online** conversation in real-time between two people or a small group. You're probably thinking, 'Hey, that sounds just like **texting** to me.' Well, it's similar. The main difference is that texting takes place over a mobile phone network while IM takes place over the internet. This makes it faster and gives you the ability to send larger **files** such as photos and videos if you want – or even take part in live video chats.

IM is usually more convenient than **emailing**. Available as **apps**, the most popular IM services are **WhatsApp**, **Facebook** Messenger, **Snapchat** and iMessage (this one just works between **Apple** devices); others include Viber, WeChat, Line, **Google** Hangouts and KiK.

See also: **MMS**, **SMS**

Intelligent personal assistant

See: **digital assistant**

Interface

To give it its full title, the 'user interface' is how you interact with a computer, TV or **mobile device**. It consists of a visual element such as the buttons, icons or menus that you see on screen, and a way to select them in order to give commands to, and control, the device. This might be a mouse, a **touchscreen** or, in the case of a **smart TV** or **set-top box**, the remote control.

Internet

Something that might seem very, very complicated is actually very, very easy to explain. The internet is simply a vast global **network** of millions of computers that communicate and share information with each other almost instantaneously. It's estimated that in 2018, over half of the world's population has access to the internet

(also known as the 'net') – over 4 billion people.
The main use of the internet is for **email** and access to the **World Wide Web**, but it's also used to transfer **files, instant messaging** and **online** discussions.
No one person, company or organisation owns or controls the internet; it's more of a concept than an actual, tangible entity – but it relies on a physical infrastructure of things such as cabling and computer **hardware** that's owned by a multitude of people: governments, commercial organisations, educational establishments, service providers and individual people. For the internet to work effectively, though, there are certain bodies that oversee and standardise things such as **domain names** and **IP addresses**.

The very first incarnation of the modern internet can be traced back to 1969 when two computers at Stanford and UCLA universities connected with each other. The first international connection took place in 1973 between London and Norway.

Internet of Things

A fridge that lets you know when you're running out of milk, a wireless thermostat that also monitors the weather or a coffee machine that knows when you're out of those expensive pods (and can even re-order them for you) . . . Welcome to the Internet of Things (IoT), a term coined to describe appliances or devices embedded with electronics, **software** and sensors that let them connect and share data over the internet. The IoT is an integral part of the concept of home automation via **smart home** systems, designed to make our lives easier.

It's been said that a new rule for the future is 'anything that can be connected, will be connected'. Will someone be able to **hack** your bank account via your toaster? Will these

devices eventually become self-aware, communicate with each other and eventually take over the world?

Who knows? But what is certain is that you can use your phone to boil the kettle remotely on your way from the bus stop to your front door.

See also: **domotics**

Internet radio

Er . . . this is radio accessed via the internet.

So what's so special about this (sometimes also called web radio) compared to listening to a conventional FM or **DAB radio**? Well, the biggest difference is that transmissions are not limited by signal strength, so as long as you have a good **WiFi** or **broadband** connection, you can listen to literally thousands of radio stations from all over the world. The second advantage is that this choice lets you listen to some very niche/specific programmes to suit your exact tastes without having to endure irrelevant playlists interspersed with banal chat. For example, there are stations that just play Italian disco classics, Eurovision Song Contest music, love songs or Bollywood hits. For classical fans there's even a station that only plays the music of Bach 24/7.

You can buy standalone internet radios if you want to listen to the radio anywhere and don't have a **mobile device**. For a full list of internet radio stations (with **hyperlinks**), visit www.internet-radio.com.

Internet streaming

See: **streaming**

Intranet

Think of this as a company or organisation's in-house internet; one that's only accessible to staff (via password access) and is isolated from the **World Wide Web**. It works in exactly the same way but just contains information relevant to the business – usually stuff that all employees need to know such as the staff handbook, holiday request and expense forms, health-and-safety procedures, the mission statement, disciplinary processes, lyrics to the company song, etc. Having all this on an intranet not only saves on paper, it means that everyone has access to the latest information (plus that information can be updated easily and quickly).

See also: **extranet**

iOS

This is **Apple**'s mobile **operating system** for its **iPhone**, **iPad** and **iPod** touch devices. To put it in context, it's the Apple equivalent of **Android** or **Windows**.

iPad

A **tablet** computer created by **Apple** in 2010 and named by *Time* magazine as one of the '50 Best Inventions of the Year'. Although there were other tablet computers available previously, the iPad is generally credited with popularising and creating the market for this type of computer. It uses the same **operating system (iOS)** as the **iPhone** and can run almost all of the iPhone apps. At the launch it was described by one analyst as 'the fastest-selling non-phone gizmo in consumer-electronics history'. Since then, over 360 million iPads have been sold.

IP address

Everything connected to the internet (not just the **hardware** but every single **website**) has a unique address to identify it called an IP address; it's a series of numbers that looks like this sort of thing: 152.131.65.325. When you're **online** it's this address that allows your device to recognise and communicate with all other devices – and find the website it's looking for.

Now, you may be thinking, 'Hold on a mo', I thought you used a **domain name** to find a website.' Well, you do – if you're human. After all, 'microsoft.com' is far easier to remember than 131.107.0.89. Computers, however, are far cleverer and use the IP address that lurks hidden behind the domain name.

IP addresses can be a complicated subject but there are really only three things you need to know about them:

* You don't need to do anything; they are assigned to your device and websites automatically.
* In essence, they're what makes the internet work for you.
* That's it.

See also: **URL, web address**

Once people know what an IP address is, most also want to know, 'Can it be used to track me?' Well, generally speaking, it can be used to show the general area where you are online, but not your actual address. If you're involved in something very dodgy, however, then a law enforcement agency can seek legal permission to contact your **ISP** and gain access to the account holder's name and address.

iPhone

One of the best-selling **smartphones** in the world, the iPhone from **Apple** was first released in 2007 and described as 'revolutionary' and a 'game changer' for the mobile phone industry. Since then there have been eleven generations of iPhone models (and probably a couple more by the time this book comes out), achieving worldwide sales of over 1 billion. Not everyone thought that the iPhone would be such a success. Steve Ballmer, the former CEO of **Microsoft**, commented when it was launched, 'There's no chance that the iPhone is going to get any significant market share. No chance.'

iPlayer

This is the BBC's **streaming** service, accessed via a **website**, a **smart TV** or by **downloading** the iPlayer **app**. As long as you're in the UK you can watch and listen to BBC TV and radio programmes live, or programmes after they've been broadcast, for up to thirty days. You can also access selected older shows from the BBC archives. And the best thing is, it's free (well, you still need a TV licence).

iPod

Apple's **MP3 player**, first released in 2001, was described as 'The Walkman of the twenty-first century'. The original model had a **hard drive** of 5 GB and allowed 1000 songs to be stored – revolutionary at the time. The iPod has gone through many iterations since then and, like most MP3 players, has largely been overtaken by the **smartphone** as a digital music player.

See also: **iTunes**, **MP3**

ISP

This is your internet service provider, a company that performs four main tasks:

1 connecting you to the internet, for which you pay a fee (this can also include telephone and digital TV packages);
2 providing you with an **email** account;
3 **hosting** your **website** (if you have one);
4 taking an interminable time to answer the phone when you have a problem.

iTunes

Pre-installed on **Apple mobile devices** (but also available for **Windows** devices), the iTunes **software** is best known as a digital music library through which you can purchase, **download**, store and organise music, but it also does the same for **e-books**, audiobooks, TV shows, films, **apps** and **podcasts**. iTunes is also used to transfer these **files** from your computer to your mobile devices, back them up, **stream internet radio** and also convert CDs into **MP3** files.

From Jailbreak *to* Junk mail

Jailbreak

This sounds unlawful, slightly dangerous and exhilarating; however, in reality it's none of these. In technology terms, jailbreaking just means **hacking** into your own **smartphone**, games console or **tablet** and installing new computer code. This allows you to override factory settings, install **apps** that haven't necessarily been approved by the manufacturer and, generally, customise how the device works. The legality of jailbreaking a device is a murky legal area basically determined by your intent – but what is 100 per cent certain is that manufacturers don't like it and it will invalidate your warranty.

So, in a nutshell, is it really worth the effort?

JPEG

When you take digital photographs they can be saved in a number of different formats. High-end cameras often save them as what's called RAW **files** where all the information recorded by the camera's sensor is retained. JPEGs (pronounced jay-pegs), on the other hand, compress this data by removing **pixels** that it determines as unimportant. This doesn't mean it will delete your sister's creepy boyfriend from that group shot; by 'unimportant' it means aspects such as sharp transitions in colour intensity and hue – changes that are almost imperceptible to the human eye. This process reduces the quality of the image but, in doing so, creates a much smaller file, which is far easier to store, **share** and display. These reasons make it the most widely used image format (you can recognise a JPEG image because the file name ends in either .jpg or .jpeg).

There's another image format called **TIFF**. Think of it like a RAW **file**, but bigger. In fact, don't really think of it at all as very few cameras have the ability to save photos in this format.

J-Pop

An abbreviation for Japanese Pop, as in Japanese pop music. The name was coined in the 1990s to distinguish homegrown pop music from foreign pop music but, confusingly, the music doesn't have to be what's accepted in the West as 'pop'. J-Pop can be poppy and upbeat, soulful or with elements of rock or thrash metal.
See also: **K-Pop**

J-Pop bands that are Big in Japan (where else?) include Babymetal, Exile the Second, Deadlift Lolita (which includes a cross-dressing Australian wrestler) and perhaps the most famous J-Pop band, AKB48, a girl band collective of 130 members that allows them to perform simultaneously at different venues as well as in their own theatre.

Junk mail

The name given to irrelevant or unsolicited **emails** sent typically indiscriminately to a large number of users. At best, this is annoying, i.e. being inundated with irrelevant advertisements (many for pills to enhance your sex life), but far more dangerous are 'get rich quick' scams, **phishing** emails, or those which link to malicious **malware** or **viruses**.
See also: **spam**

From Kardashians *to* K-Pop

Kardashians

America's Number One Family since the Kennedys, it's a well-worn and lazy cliché to say that the Kardashians are only famous for being famous.

But they are.

They don't sing, play instruments, act, write literature, compose poetry, play sports or fight world hunger or social injustice. Their particular talent is letting cameras follow them around while they show off how opulent they are, while also having inter-family squabbles (especially for the camera).

The TV series *Keeping Up with the Kardashians* showcases the extended family, in particular the relationships between the mother Kris, her three daughters Kim (who is married to singer Kanye West and has around 60 million **Twitter followers**), Khloé and Kourtney and their half-sisters Kendall and Kylie Jenner. It's been on air since 2007 and has won several television awards.

Five things you may not know about the Kardashians

1 Khloé Kardashian made a video about how to put cookies in a jar and it has so far received over 3.6 million views on **YouTube**.

2 Kourtney, Kim and Khloé once rented their mother a monkey for a few days because 'she had a syndrome where she missed children in the house'.

3 Kris claimed that having a neck-lift was a 'transcendental, life-changing experience'.

4 Kim said that, 'I hate it when women wear the wrong foundation colour. It might be the worst thing on the planet when they wear their make-up too light.'

5 And also, 'A Rolls-Royce is the best car if you have kids, spacewise.'

Kickstarter

One of the best-known **crowdfunding** companies, formed in the US in 2009, and focusing on creative projects.

Kindle

Amazon's **e-reader** brand, introduced in November 2007. The first model was so popular that it sold out in five and a half hours and remained out of stock for five months.

See also: **e-book**, **e-ink**

The name 'Kindle' was chosen for its meaning to light a fire; a metaphor for reading stimulating intellectual excitement.

K-Pop

The **J-Pop** of South Korea which, like its Japanese counterpart, is influenced by a diverse range of musical styles including Western pop music, reggae, R&B, gospel, hip-hop, electronic dance and even folk. K-Pop achieved global recognition with the colossal success of 'Gangnam Style' by the singer Psy in 2012. Today, one of the biggest K-Pop bands is BTS, a seven-member boy band whose 2018 album reached number one in the US. They were the most **tweeted**-about celebrities in 2017, being 'liked' or retweeted over half a billion times worldwide.

The video for 'Gangnam Style' was the first **YouTube** video to reach one billion views (the figure is currently over 3.15 billion). United National Secretary General Ban Ki-moon hailed the song as a 'force for world peace'.

Quite.

From Laptop *to* Logic bomb

Laptop

A portable battery-powered computer that traditionally uses what's known as a 'clamshell' design with a screen in the lid, a keyboard and **touchpad** in the base – and a hinge in the middle. Sometimes also called 'Notebook computers', laptops are, to some degree, being replaced by **Chromebooks** and **tablets** which are lighter and have a longer battery life. Laptops, however, still have the advantage of being more powerful and having more storage for data.

The first laptop dates back to 1982 and was 2 inches thick . . . eighteen times thicker than the thinnest part of some current laptops.

---

lastminute.com

Set up in 1998 to cater for people who were too busy, lazy or absent-minded to plan ahead and book flights, hotel rooms or whole holidays, lastminute.com was one of the first **online** retailers in the UK and caught the attention of the public. Over twenty years later it's still very much in the public consciousness – so much so that the company name has become a catchphrase for anything left to the, well, last minute. Today, the **website** also offers theatre tickets, car hire and days out; it works by getting what's called 'excess inventory' from suppliers, e.g. unsold airline tickets, hotel rooms, etc., and offering these at a reduced rate. The later you leave it, the better the discounts (in most cases) since these items won't exist a day later if left unused, and therefore have nil value to the companies selling them, e.g. the airlines or hotel operators. If you really can wait to the last minute before you book a holiday and are flexible about the destination, there are some really good bargains to be had.

LGBT

A term used to discuss the Lesbian, Gay, Bisexual and Transgender community, a replacement for the old term 'gay community' that expresses more of a diversity of sexuality. Sometimes you'll see LGBTQI, the Q representing those who identify themselves as Queer or Questioning their sexual identity, and the I representing those who are Intersex (the term for people previously known as hermaphrodites).

To make it even more inclusive, however, other initials have been added so that the expanded term can be LGBTTQQIAAP (Lesbian, Gay, Bisexual, Transgender, Transsexual, Queer, Questioning, Intersex, Asexual, Ally, Pansexual).

To make it simple, the term LGBT+ is sometimes used.

NB An 'Ally' can be someone who isn't LGBT+ but who supports the community, or someone within the community who supports those with a different sexuality.

Life coach

In essence a life coach is a combination of a mentor, motivational speaker, consultant and therapist, and maybe a bit of an agony-aunt thrown in for good measure. Their role is to understand what you want from life – personally or professionally – and then help you achieve that. It could be that job promotion, balancing your work and home life, getting out of a rut, a career change, building self-worth, improving a relationship . . . anything where there's a gap between where you are now and where you want to be. The life coach's job is to unlock your potential and close that gap. Oh yes . . . and charge you a lot of money in the process.

Lightning cable

If you own a 2012 or later **iPad** or **iPhone** you'll know what this is; it's the ultra-slim connector with eight tiny 'pins' that charges your device. **Apple** introduced it to allow its devices to be made slimmer; it's not compatible with any other device.

Link

See: **hyperlink**

LinkedIn

In simple terms, think of this as **Facebook** for business. It's a **social media website** aimed at connecting you with people you know professionally – previous, current or potential colleagues. LinkedIn currently has over 467 million members worldwide and is available for free or with more features available for a monthly subscription. It's great for networking, listing/looking for job opportunities and garnering feelings of resentment when you discover that the spotty junior you once managed is now the CEO of a FTSE 100 company.

Live streaming

This means watching and listening to video/audio in real-time over the internet, rather than **downloading** it to your computer or **mobile device** and watching/listening to it later. As its name suggests, what you see and hear is being broadcast live, which makes it popular for one-off events such as concerts or festivals, and reality TV shows such as *Big Brother*. Live streaming helps to create the feeling of 'being there' plus, of course, it also gives you the chance to hear any swearing before it's bleeped.
See also: **simulcast**, **streaming**, **streaming stick**, **video on demand**

Logic bomb

This is **malware** that's programmed to be triggered by a specific event. The event could be a certain date/time, for example, or when certain **files** are opened. There's a history of logic bombs being set by disgruntled employees as a way to get revenge after being let go or fired. They can be designed to be malicious and do serious damage to a company, e.g. by deleting critical files – or just be mischievous, e.g. by **emailing** details of every member of staff's salary throughout the company.
See also: **virus**

From Malware *to* MySpace

Malware

A contraction of *malicious software*, malware is harmful **software** that's specifically designed to damage data and/or computers – or gain unauthorised access to them. It can take many different forms, the most common being **viruses**, but there are also nasty things known as Trojans, ransomware, adware, **spyware**, **logic bombs**, rootkit and worms. Malware can end up on your device from infected **files**, usually **downloaded** from **spam emails**. Knowing the type of malware isn't that important. Knowing how to protect your computer or **mobile device** is – and that's by being wary about unsolicited emails and by installing the latest anti-virus software and/or anti-malware software.

Manga

Just as **anime** has come to mean (in the West) all animation from Japan, manga (pronounced man-gah) is the term used to describe Japanese comic books and graphic novels; the word means comic or cartooning. Manga characters can be recognised by their large eyes and small mouths, brightly coloured hair and display of exaggerated emotions. Like anime, and unlike Western comics, manga is mainstream in Japan and aimed at both adults and children.

Unlike Western comics, most manga are black and white (as this allows for faster production) and are the size of small books that are collected in volumes. It's not only the pages that are read from right to left, the panels and text are read the same way.

#MCM

Man Crush Monday; an opportunity to use that **hashtag** and **post** a photo of a good-looking guy on **Twitter** or **Instagram**, on the basis that he's your crush for the whole day. Designed to brighten the lives of everyone who likes looking at attractive men.

Meme

Grumpy Cat, Doge, Hitler ranting about something, Bad Luck Brian, Leonardo DiCaprio toasting something, LOLcats, **Rickrolling**, Meryl Streep singing, Distracted Boyfriend, Kylo Ren's high-waisted pants, Jennifer Garner clapping and stopping at the Oscars . . . you've probably seen them and maybe even **shared** them. These are all examples of memes (rhymes with 'teams'): an image, video, word or phrase (often intentionally misspelt or with bad grammar) that is spread from person to person via the internet for entertainment value. It usually takes the form of a **post** or a **hyperlink** to a **website** on **social media** or in an **email**.

It's this mass sharing (sometimes worldwide) that gives the meme its status.

The thing with memes is that they change all the time. What's funny one month (dancing hamsters or bizarre Chuck Norris facts) is sooooo out of fashion the next.

See also: **viral**

The word meme was first coined in 1976 devised by the evolutionary biologist Richard Dawkins. It comes from the Greek word *mimema*, which means 'imitated thing'. Dawkins described memes as a way for people to transmit social memories and cultural ideas to each other.

Menu

In the same way that a restaurant menu presents you with a selection of choices for food and drink, a **software/ program/app** menu is a list of options that help you decide what you want to do. A menu on a word-processing program would, for example, give you these choices: create a new document, open a previous document, save a document, save that same document but under a different name, close a document or print it.

Clicking on a button or link makes the menu visible – usually as what's called a 'pop-up menu' or a 'dropdown menu' (the names are self-explanatory!).

Meta

You're watching a movie and one of the characters makes a joke about imagining he was in a movie being watched by an audience . . . that's an example of something meta. So is 'A priest, a rabbi and a horse walk into a bar. The horse looks around and says, "I'm in the wrong joke!"' Basically meta is used to describe something that's is self-referencing . . . normally used for comedic effect.

Microblogging

These are very short **blogs**, usually consisting of a message of just a few sentences, a **hyperlink**, an image or a video, but rather than appearing **online** as a permanent blog, they are **posted** and **shared** across **social media**. **Twitter**, **Tumblr** and **Facebook** are considered microblogging **platforms**.

Microsoft

A hugely successful US multinational technology company co-founded in 1975 by Bill Gates and best known for its MS-DOS computer **operating system** in the 1980s, which later evolved into **Windows**, and its Office suite of business **software** (which includes the ubiquitous

Microsoft **Word**). Microsoft has since diversified from the **software** market (although it still remains the world's largest software company) and now has its fingers in many different technology pies. Since the 1990s the company has launched the very successful Xbox video game console and the Surface range of **tablet** computers (as well as a less successful range of Windows phones and Zune **MP3 players**). Also in Microsoft's extensive product portfolio is the **Bing search engine**, the Outlook **webmail** service, the OneDrive **Cloud** computing service and the web **browser** Internet Explorer. Microsoft acquired the mobile phone division of Nokia in 2014 and is also the parent company of **LinkedIn** and **Skype**.

Unlike rivals **Apple**, the origin of the company name isn't hard to guess; it's a combination of the words 'microcomputer' (or 'microprocessor') and 'software'.

Millennials

Also known as 'Generation Y', these are people born between the early 1980s and the early 2000s. While it's irresponsible to generalise a whole section of society, Millennials tend to be recognised by these ten characteristics:

1 Tendency to work in the media/creative industries
2 Have a liberal or left-wing attitude to politics and economics
3 Value diversity and are very politically correct
4 Less patriotic but more globally minded
5 Obsessive about **social media**
6 And **avocado toast**
7 Have a 'want it all and want it now' attitude
8 Think they know how to run the world
9 Act overly cool and entitled
10 You dislike them

See also: **selfie**

Mindfulness

For many, the relentless and exhausting pressures of a frantic modern life can lead to feelings of stress, anxiety and even depression. Some people have a way of dealing with, and controlling, these feelings of angst. It's called Prosecco. There's also another, more healthy and spiritual method that's definitely **on-trend**; this one is known as mindfulness.

In simple terms, it's a form of meditation where you concentrate on your breath as it flows gently in and out of your body. Focusing on each breath this way allows you to consider your thoughts as they arise in your mind, and observe them with what practitioners of mindfulness call 'friendly curiosity'. This way you can catch negative thoughts before they take hold – and begin the process of putting yourself back in control of your life.

Over time mindfulness is said to improve your mood and bring about long-term improvements in your general happiness and sense of wellbeing. Mind you, some people claim that same effect from Prosecco.

See also: **wellness**

MMS

Multimedia Messaging Service is a form of **text messaging**; one that allows you to send not just text but photos, videos and other **files** over the mobile phone network. Unlike regular text messages that are limited to 160 characters, an MMS message can contain 1200.

See also: **SMS**

Mobile broadband

If you don't have, or don't want, a telephone line and the corresponding line rental costs but do want the advantage of **broadband** wherever you are, then mobile broadband is the answer. This uses the mobile phone network to connect to the internet but you don't need a mobile phone

– just a **modem dongle** which plugs into a **USB** port on your computer.

It sounds great having access to the internet anywhere and at any time but, remember, you're at the mercy of the mobile phone signal. This can mean bad connections or slower **download** speeds at peak times.

Mobile device

Sometimes just called 'device', this is a general term for a **smartphone** or a hand-held computer (typically **tablets** and **e-readers**) with internet connectivity via **WiFi** or a mobile phone network. Nowadays mobile devices can do most things that were traditionally carried out by desktop computers or **laptops**.

Modem

Usually built into your **broadband router**, in simple terms a modem is the device that allows a computer to send and receive data to and from your **ISP** over a telephone line or a fibre-optic cable. That doesn't sound very important but, believe me, oh yes it is.

In short, it's the modem that provides you with access to the internet.

Moderator

Most **chat rooms** and **forums** usually have a set of rules to prohibit things such as making offensive comments or using bad language. Volunteers called moderators lurk in the background, welcoming or helping new members – but are mainly there to keep an eye out for undesirable behaviour, with the power to bar someone from the chat room/forum. Depending on the rules of the chat room/ forum and the 'offence' committed, this ban might be for a day, a few days or forever.

MP3

First there were wax recording cylinders, then records, then reel-to-reel tapes, then eight-tracks, then cassettes, then CDs . . . Now it's the turn of the MP3 as the flavour-of-the-month music format. It's a digital music format that can compress songs and enable you to fit hundreds (or even thousands) of them on your **mobile device** or computer with minimal loss of quality. That means that even a small (1 GB) **flash drive** can hold 250 Celine Dion or Westlife songs.

(If you must.)

See also: **MP3 player**

Many people convert their CD collection to MP3 files so they can play them on their computer or transfer them on to a mobile device. You can also purchase tracks in MP3 format for about £1 each via services such as **iTunes**, **Google Play**, **Spotify** or **Amazon**.

MP3 player

This could be a **smartphone**, a **tablet** or a standalone device such as **Apple**'s **iPod**. It's simply a device that can play **MP3 files**, which means you can store and play music, audiobooks and **podcasts**. An MP3 player basically consists of a memory chip for storing the files, a battery to power it, controls and headphone socket (and/or **Bluetooth** facility to work with wireless headphones).

Multiplayer gaming

Video gaming is usually a solitary activity; it's man (or woman) against machine – where the machine is the games console, which reacts with a series of pre-programmed responses. On the other hand, a multiplayer game is far more random and unpredictable – and therefore exciting – since it allows you to play the same game 'live' with other players.

Players are connected with each other over the internet and can play against one another or can cooperate and play against other players, or the console itself. The variations possible depend on the actual game being played. Multi-player games tend to support between five and fifty players simultaneously while Massively Multiplayer Online (MMO) games can support several thousand players who can join and leave the game any time they want.

Mumsnet

One of the most popular – and some say influential – parenting **websites**, which was launched in 2000 and now has 12 million unique visitors each month. One of the best-loved aspects of the website is the Mumsnet **forums**, which host discussions on anything from serious medical or childcare issues to users complaining that their husband spends too much time on the toilet.
Visit: www.mumsnet.com

Music streaming

Imagine having the ability to listen to any song (or 'track' as the youngsters call them these days) by the Bee Gees, Showaddywaddy or Herman's Hermits whenever or wherever you want! You can, thanks to music streaming (also called 'music on demand') – the ability to listen to music in real-time over the internet, rather than having to **download** it to your computer or **mobile device** and listen to it later.

Literally millions of tracks are provided by what's called music streaming services including **Amazon** Music Unlimited, **Apple** Music, Deezer, **Google Play** Music, Napster, Pandora and **Spotify**.

Streaming is a relatively recent development, made possible by faster **broadband** and **WiFi** connections, and is available free or via a monthly subscription (which brings other benefits such as more choice of music and better sound quality).

Musk, Elon

Not a new deodorant or aftershave, Elon Musk is a South African-born entrepreneur who founded an **online** payment company, which eventually became **PayPal**, and which was bought by **eBay** for $1.5 billion in 2002. Today he is best known for founding SpaceX, a 'space transport services' company (i.e. it makes rockets) and as co-founder of **Tesla**, best known for its electric cars. In February 2018 he combined both of these business interests by launching his own Tesla car into orbit.

Because he could.

MySpace

From 2004 to 2010 this was the largest **social media** site in the world, creating one of the most popular **platforms** for new bands and artists to be heard and grow their fan base. In 2008, it was overtaken by **Facebook** (which was founded a few months later in 2004) and since then it has declined massively in popularity and membership. Many people who remember MySpace as one of the first social media sites are surprised to hear it's even still in existence, but it is and thriving – in its own way. Today, it carries a heavy amount of entertainment news but is still a **social**

networking site at heart with the ability to create a really personalised **profile** page where you can add music, blogs and photos/videos.
Visit: www.myspace.com

Artists that launched their career/were discovered on MySpace include the Arctic Monkeys, Lily Allen, Calvin Harris, Panic! at the Disco and Kate Nash (you may have heard of some of these acts).

From Nanobot *to* Newsgroup

Nanobot

A teeny, weeny 'robot'. These bots are so small that they operate at a microscopic level (about the size of a red blood cell), and so small that they can travel in your bloodstream, conduct health scans and eradicate bacteria, cancers and viruses. The first successful nanobots were engineered DNA used in March 2018 to shrink tumours in mice. The military is also looking at the technology – its interest includes a way to accelerate the healing of injured soldiers and the repair of equipment on the battlefield, as well as creating eavesdropping devices that are virtually undetectable.

See also: **nano technology**

Nano technology

Also called nanoscience, this term refers to any technology that involves studying and working with things that are tiny. In fact, tiny is a gross overstatement. These things are between 1 and 100 nanometres in size. A nanometre is one billionth of a metre . . . it's hard to imagine anything that small but, to put it in perspective, a single nanometre is one hundred-thousandth of the thickness of a page in this book – or to express it another way, it's how far your fingernail grows in one second.

So what do nanoscientists do (apart from looking at their fingernails and feeling smug)? They spend a lot of their time looking down microscopes and manipulating individual atoms and molecules in order to create brand new materials or improve existing ones. Applications of nanotechnology are extensive and eclectic; from creating bandages that speed up the healing process, clothes that are more stain resistant or golf balls that fly straighter.

See also: **nanobot**

Netbook

The generic name given to a genre of small, lightweight no-frills **laptop** that was introduced in 2007. Four years later netbooks had been overtaken in popularity by **tablet** computers. Most manufacturers stopped making netbooks in 2012; the modern incarnation is the **Chromebook**.

Netflix

An American entertainment company that provides **video on demand**, and which has recently expanded into film and television production for which it has won an assortment of Academy, Emmy and Golden Globe awards. Netflix **streams** video in over 190 countries to nearly 118 million subscribers; for a monthly fee you can access a wide range of titles, from **box sets** of TV series to old and new (well, recent) movies across a huge range of genres. Netflix is a direct rival to **Amazon** Prime Video.

Five things you may not know about Netflix

1 It started life in 1998 as an **online** DVD rental store; it posted the DVDs to you and you posted them back afterwards. It still offers this service in the US.

2 Netflix is available in all countries apart from Mainland China, Crimea, North Korea and Syria.

3 The current top-three shows that subscribers **binge-watch** first are *Orange Is the New Black*, *Breaking Bad* and *The Walking Dead*.

4 It knows what subscribers watch; it reported that a single person in England watched Jerry Seinfeld's animated *Bee Movie* 357 times in 2017. Someone else watched *Pirates of the Caribbean: The Curse of the Black Pearl* 365 days in a row.

5 Netflix was in financial difficulty in 2000 and offered Blockbuster the chance to buy it for $50 million. Blockbuster declined. Netflix is now worth more than $100 billion. Blockbuster is now worth, er, well . . .

Netiquette

Just as etiquette is a code of acceptable behaviour in society, netiquette is a term coined for a code of polite behaviour on the internet. This covers all aspects of **online** communication including **email**, **social media**, **chat rooms/forums**, **multiplayer gaming** – basically anywhere you're interacting with other people (usually complete strangers) via your keyboard.

The code isn't official; it's a series of generally accepted conventions designed to make online communication constructive and enjoyable. Breaking any of the 'rules' won't prompt a visit by the Internet Police, but sticking to them will show others that you respect their views and your mother brought you up well (even though your mother might not know what the internet is).

Twelve netiquette rules

1 Don't **post** inciting, abusive or offensive comments online (i.e. **flaming**).

2 Respect others' privacy; don't quote or forward someone's email address or phone number without their permission.

3 Don't be a potty mouth (in the rare occasions where you need to swear to make a point, use aster*sks).

4 Don't type in UPPERCASE LETTERS. IT'S HARD TO READ AND LOOKS LIKE YOU'RE SHOUTING!

5 Don't bombard people with massive amounts of unsolicited email (i.e. **spam**).

6 If you're playing online games, show good sportsmanship; don't be a sore loser or an arrogant winner.

7 When posting or making comments, stick to the topic in question.
8 Don't overuse **emoticons** or **emojis**. You're not a twelve-year-old girl (unless you are, of course, in which case it's acceptable and expected).
9 Don't steal content and pass it off as your own. If you use someone else's images or text, give them credit BUT . . .
10 Don't post material (images, large amounts of text) that is copyrighted (which probably applies to most of the images you find online). The rights holder could take legal action against you and/or the **website**/forum in question.
11 Respect the views of a forum or chat room's **moderator**.
12 Be patient and courteous with inexperienced users. You were once new to the whole online environment – and probably still are.

Net rep

Short for 'internet reputation', and also known as your '**online** reputation'. Like all reputations, it's important to have a good one, but it's even more important online where a business can find itself up against a lot of unknown, faceless competitors – many offering (unsubstantiated) lower prices or promising better service. Or both. Having a good net rep can give a company status and kudos – and can also help to overcome any negative perceptions.

So what makes up your net rep? Well, it's the **content** of your **website** or your **blog**, your **social media posts**, your interaction with customers, online reviews such as those on **Trustpilot** – in essence, anything and everything online.

See also: **digital footprint**

If you're a business it's good to **Google** yourself regularly to see what others see when they do the same thing.

Network

A computer **network** is simply two or more computers connected to each other via cables or wirelessly in order to 'talk to one another' and **share** information. The world's largest network is the internet.

Newsgroup

Newsgroups are web-based discussion groups covering a wide range of different topics – and by wide range, I mean, for every subject imaginable. There's estimated to be over 100,000 newsgroups with millions of users covering topics from the general, e.g. fine arts, taxation and atheism, to the much more niche such as *The X-Files*, Abba and working with glass.

Although newsgroups pre-date the **World Wide Web** (they began way back in 1979 as a way for researchers at universities and technology companies to exchange messages and **files**), they are similar to internet **forums**; the main difference is that they are accessed via a monthly subscription on something called Usenet (a global system of discussion groups), rather than the World Wide Web, and many of them operate without a **moderator**.

Visit: http://usenet-deluxe.com/en

See also: **chat rooms**

Usenet was the first internet community and the place where Tim Berners-Lee announced the launch of the World Wide Web in 1989.

From Oculus Rift *to* OSX

Oculus Rift

Not one of the Transformers but one of the first **virtual reality** headsets (and incidentally, funded by a **Kickstarter** campaign in 2012).

Offline

The opposite of being **online**.

Online

Being online simply means that a person, a computer or other device is connected to the internet or any other computer **network**.

Online dating

Dating (or, as it used to be known, 'courting') is so different nowadays. It used to be far easier. Back in the day the person you'd end up marrying used to be the boy or girl next door, someone in the youth club or church, or someone introduced by a mutual friend. Then there was an established structure to the process that both parties understood. First dates might include a trip to the movies, a picnic, a nice stroll in the park or even a fancy meal, then taking your date home at a reasonable time. Perhaps if it

went well, you might get a quick kiss and a promise to see you again. If there was mutual attraction the dates would progress to 'going steady', a time to really get to know your boyfriend/girlfriend and be introduced to their circle of friends and their parents.

Nowadays no one knows their neighbours, goes to youth clubs or church and, to be honest, no one has the time these days to go on endless dates to get to know someone. In this modern world, it's all about convenience and instant gratification and that's where online dating comes in.

Most online dating is conducted via dating **apps**, although some also exist as websites (some are free and some you pay for). Although there's a wide choice of dating apps depending on your age, sexual orientation, religion and whether you have any, ahem, 'niche' interests, they all work more or less the same way.

You open an account and create a **profile**, then set your search parameters (gender, age range, location, etc.). You're then presented with a selection of potential dates that supposedly match your search criteria (but which, to be honest, rarely do).

Different dating apps have different ways you can express interest in a person but the better apps (and by 'better' I mean the ones where you're less likely to be contacted out of the blue by creeps/bunny boilers) only permit contact if there's a mutual interest. That's the time you can then establish your true compatibility by messaging them (or as it used to be known, 'talking').

NB When you meet someone using a dating app, remember that it wasn't fate that brought you together. It was an **algorithm**.

See also: **catfish**, **Grindr**, **swipe left/right**, **Tinder**

Some useful online dating terms

Breadcrumbing: leading someone on without any intention of getting serious with them.

Caspering: a nicer way to let someone down than ghosting (named after the friendly ghost). This involves telling someone how you really feel before disappearing from their digital lives.
Ghosting: abruptly cutting off all contact with someone with whom you've been communicating. Think of it as splitting up before you even get to the stage of splitting up.
Marleying: when an ex contacts you via **social media** at Christmas out of nowhere.

Online gambling

As its name suggests, this is a way to lose large amounts of money without leaving the comfort of your home. Also called internet gambling, players can bet on actual sporting events but the biggest innovation is playing casino games against the 'house' (i.e. the **website**) or against other players in real-time. These games are the same as the real-world ones, the most common being poker (various versions), roulette, blackjack, baccarat and, of course, bingo. Due to the virtual nature of this form of gambling, there's no way for players to be sure that the various gambling websites are fair, i.e. the correct amount of randomness in a game. The best bet (see what I did there?) is to do your research, read reviews and talk to other players in one of the many **forums** for online gambling in general, or specific games in particular, and then only play gambling sites with the best reputation.

Signs that an online gambling website might be, well, 'shady' or in financial trouble include the customer service or support team giving you the runaround when you have a query, delays in paying your winnings or, worse still, coming up with a litany of excuses as to why they don't have to pay you.

On-trend

The modern, acceptable way of saying 'trendy'.

Operating system

Windows, **iOS**, **OSX**, **Android**, **Chrome** . . . these are what are known as operating systems – they're basically pre-installed powerful **software programs** that make your computer or **mobile device** work. It's the operating system that controls the general operation of the device and provides an easy and intuitive way to use it. Without its operating system, a computer wouldn't be a computer. It would just be a plastic case, a screen and a load of electronic components.

OSX

Apple's own **operating system** for its range of desktop computers and **laptops**.

From Passcode *to* Program

Passcode

For security reasons you may be asked to enter this to allow you to access your **mobile device** or **download** an **app**. It's just the same as a password, except it involves numbers not letters ('passnumber' just sounds odd). In the future, the more widespread use of **facial recognition** will probably make passcodes (and passwords) obsolete.

PayPal

Wouldn't it be good to have a payment system that was easy to use, fast and secure – and didn't require you to enter your credit card details each time or remember where you left your cheque book. There is and it's called PayPal – perhaps best known as a way to make payments on **eBay** and on a huge selection of other online retail **websites** – as well as providing a way to send money to anyone with an email address.

It's a worldwide online payments system that links your credit card or bank account to your PayPal account. The system remembers your details so each time you make a payment you just need the password for your account; your bank details aren't shared with sellers. Any money received goes into your PayPal account; you then transfer it to your bank account. Of course, none of this happens for free. PayPal charges sellers a range of fees for transactions.

The service was launched in 1999 but began life with the name Confinity, becoming PayPal in 2001. One of its founders was Elon **Musk**.

Paywall

This is what **website** owners put up between their website and your computer to prevent free access to their **content**; only those people paying a subscription are 'allowed in'. In the face of a massive decline in revenue from advertising, paywalls tend to be found on newspaper publishers' websites. There are two types of paywall. 'Hard paywalls' are where you have to pay to access any **online content**. This is a risky strategy as there's a high likelihood that users will go elsewhere for their content (a US survey in 2017 showed that 82 per cent of respondents who read their news online would rather find a free alternative than pay for their preferred website). On the other hand, a good compromise is a 'soft paywall', which grants access on a limited basis, and then asks for payment in order to continue.

PDA

Nowadays delivery drivers, utility company meter readers, the military, medical staff, waiters and parking wardens are among the specialist groups that still use PDAs. In this sense, PDA does not mean Public Display of Affection (especially not with parking wardens); it stands for Personal Digital Assistant. These are small, hand-held computers that were traditionally used for recording notes and tasks, and as diaries, address books and calculators – although there were also many applications available for specific business purposes. The introduction of **smartphones** in the early 2000s meant most PDAs for personal use ended up in that great technology skip in the sky.

PDF

Developed by Adobe Systems, this is a Portable Document Format **file**. It's commonly used as a format for **downloadable** brochures, product manuals, scanned documents and flyers. The great thing about PDFs is that they don't rely on the **software** that created them or any particular **operating system** or **hardware**. That means that as long as you have Adobe Acrobat Reader on your **device** (available free from https://get.adobe.com), you can open any PDF and it will look the same no matter what type of computer or device it's viewed on.

Peripheral

Peripherals are ancillary devices connected to your computer or device that add functionality to it. Examples of peripherals are scanners, printers, **webcams**, a keyboard and a mouse. They can be connected physically via a cable or wirelessly via **WiFi** or **Bluetooth**.
See also: **hardware**

Periscope

This is an **app** that allows you to **stream** live video direct from your **smartphone** or **tablet**. It's owned by **Twitter** and integrated into the Twitter app so you can instantly **share** your 'live broadcast' with any of your **followers** (and also via **Facebook**). Like **Snapchat**, videos shared via Periscope are ephemeral; if you miss the live broadcast you can watch a replay up to twenty-four hours later but, after that, the video is removed from the app. The idea is that you can capture the 'immediacy' of the atmosphere whether, for example, it's watching your favourite football team or band, hang-gliding, going on a rally or march, surfing or walking through a bustling Tokyo. In October 2016 a UK mother planned to stream the birth of her fourth child on Periscope; however, it was reported that labour pains got the better of her and she smashed her phone 'during a particularly taxing contraction'.

Personal trainer

Sometimes it's not enough to go to the gym and exercise on your own. Sometimes you need someone for an hour or two each week who'll put you through such pain and despair that you'd have thought that the especially fit young man or woman standing inches from your face was rather Torquemada or Vlad the Impaler. That said, they're just

doing their job. In basic terms, the role of a personal trainer (referred to as a PT) is threefold: to design a tailored personal fitness regime to meet your goals (it could be something general such as losing weight or gaining muscle, or something specific such as 'lifting and shaping your lower body assets'); to motivate you while you're trying to meet those goals; and, importantly, to monitor how well you're doing. Most gyms offer a personal training facility.

Phishing

You know that **email** you received from Nigeria saying you are a distant relative of Prince M'Dingi Lotsomoney and are in line to inherit his $82,000,000 fortune following his tragic and untimely death – and all you need to do is send your passport number and bank details to confirm your identity? Well, that's an example of phishing: a **cybercrime** whereby the perpetrators try to trick someone into providing confidential information (usually passwords or financial information) in order that someone can con you out of large sums of money.

While many people are not fooled by this sort of scam (although it's surprising how many are), phishing emails can come from what appear to be reputable organisations and people with whom you already have a relationship, which makes them very difficult to spot. Some phishing takes place over the phone or via **text messages** but most takes place over the internet. In these cases, the email or **website** addresses used are usually very similar to the genuine ones . . . but there's a slight discrepancy.

If you're in any doubt over the authenticity of an email, contact the company in question via the contact details you usually use – and never, ever, ever, ever give out any passwords.

Visit: www.phishing.org

See also: **hacking**

Photoshop

Forget an exfoliating scrub, a non-greasy, hydrating face oil or a detox clay mask . . . the must-have beauty product for models nowadays is Photoshop. A product that's become a verb, Photoshop is industry-standard image-editing **software** created by the company Adobe, but the term has now become generic for any software that does the same job. And by image editing, we really mean image manipulating. Teeth that are too crooked, arms that are too fat, freckles that are too, well, freckly . . . all can be changed with Photoshop. It's difficult to remember a time when models had to be naturally attractive.

See also: **Photoshop fail**

Photoshop fail

The best **Photoshop** technique relies on having the eye of an artist and the skill of a surgeon; the results should be subtle so that no one suspects that any digital jiggery-pokery has taken place. Any image where the manipulation is obvious (sometimes stupidly so) is a Photoshop fail. These are more evident in work that's been done on people shots; obvious clues are waists or legs that are impossibly thin or long, or hips that are thinner than thighs. Some genuine Photoshop fails have even resulted in celebrities and models having two right feet, or losing fingers, arms, knees and even belly buttons.

Pinterest

Described as a 'visual **social network**', this is a sort of **online** pinboard – a place where you can collect (and **share**) images of things you're interested in. Each different collection or theme is called a 'board' and each image you add is a 'pin'. Your board could contain pudding recipes, motivational quotes, South American reptiles, wooden furniture, homemade jewellery, designer shoes, make-up tips, modern architecture, wedding venues, retro-styled motorcycles – anything as long as it's visual.

Like most **social media** sites, you can follow the boards of friends (comment and 'like' their pins), re-pin images to your own boards and share their pins on other sites such as **Facebook** or **Twitter**. According to the site itself, 'Pinterest is where people discover new ideas and find inspiration to do the things they love!' (Pinterest's exclamation mark.)

Visit: www.pinterest.co.uk

Pixel

All digital images (photos, illustrations, graphics, etc.) are made up of loads and loads of different coloured tiny dots called pixels. These are packed really tightly next to one another so, when viewed on a computer screen, the individual dots are not noticeable. The greater the number of pixels in an image, the smoother, sharper and larger the image can appear on a screen. If you try and enlarge an image that has a low number of pixels, you'll see it quickly loses its sharpness, a result of its original individual dots becoming more visible. In this case, you'd say the image was 'pixelated' (less technical people would use the word 'fuzzy').

Images might be described as 4032 × 3024. This means they are 4032 pixels wide and 3024 pixels high giving an overall size of 12,192,768 pixels (or 12 megapixels).

Platform

This is used to describe an **operating system** as in, 'This **software** only works on the **Windows** platform' (you can, of course, just say, 'This software only works with Windows' but including the word platform makes you sound more techy).* Nowadays though, platform is more commonly used to refer to different types of **social media** so you might say, 'I'm using the platforms **Twitter** and **Instagram**.'

*Next thing you know, you'll be using the word **algorithm** with gay abandon.

Plug and play

If you see this phrase (sometimes written as 'plug 'n' play', 'plug N play' or PnP) in connection with a piece of computer **hardware** then it's good news. What it means is that the **operating system** on your computer or **mobile device** will automatically detect and configure (i.e. set up) the scanner, printer, **webcam**, keyboard, mouse – or whatever **peripheral** device you're adding. Before PnP, it was a dark period in computing involving configuring things using confusing initials such as the DMA, IRQ and I/O addresses.

Believe me, when it comes to computing you want to shy away from anything that involves the word 'configure'.

Podcast

The 'pod' bit of the name is borrowed from **Apple**'s **iPod**, its innovative **MP3 player**, while the 'cast' alludes to the word 'broadcast'. With that in mind, think of a podcast as a radio show that's available via the internet, although some are more frequent, and some are generated specifically as **online** podcasts. You can **download** the podcast, usually for free, and play it on an MP3 player (not exclusively; you can also listen on your computer, phone or **tablet**). That means that you can choose exactly where and when you listen – and not be a slave to the whims of the broadcaster.

You can find podcasts on Apple's **iTunes** store or **websites/apps** such as Stitcher, Earwolf or RadioPublic.

Figures vary considerably as to how many active podcasts there are . . . but it's safe to say over 250,000 – and maybe up to 500,000 – on topics as diverse as movie reviews, current affairs, feminism, *Star Trek* or true crime, to very niche podcasts such as those about brewing beer at home, 1990s football matches, pens (yes, you read that correctly) and even one with the self-explanatory title, 'Denzel Washington Is The Greatest Actor Of All Time Period'.

Pokémon GO

Introduced in 2016 to almost cultish levels of interest, this is an **augmented reality** (AR) game **app** developed by Nintendo and designed to be played on **Android** and **iOS mobile devices**. It uses your phone or **tablet**'s **GPS** signal and its camera to show a menagerie of over 200 virtual creatures (called Pokémon) on screen superimposed on your actual surroundings, as if they were in the real world. Players (called 'trainers') would walk around an area until they're alerted to the presence of a Pokémon and then battle, capture and train it to gain points.

The game attracted praise for its game play, the way it popularised AR technology and how it promoted physical activity. It also attracted controversy by the way it encouraged users to capture Pokémon at locations deemed insensitive, unsuitable or dangerous. These have included cemeteries and memorials (including the United States Holocaust Memorial Museum and the Hiroshima Peace Park), places of religious worship, a location set in the Korean Demilitarised Zone and even minefields left over from the 1990s Bosnian War – although some of these were later deleted from the game.

Pokémon GO became one of the most popular apps in 2016 and was **downloaded** more than 500 million times worldwide that year, although its popularity had declined by the end of the year.

Due to the controversy resulting from crowds of players convening on 'unsuitable' locations causing massive disruption, some governments banned the game from certain areas. In some countries the game was described as 'demonic' and Russia expressed concern that foreign governments might be using the app to gather sensitive information. The game has also been associated with a number of injuries and deaths resulting from players chasing Pokémon into busy roads or playing it while driving.

Pop-up

You're browsing the web, maybe reading an interesting feature titled 'Could YOU be a royal bride?' or watching a video of a cat playing the piano, when an advertisement suddenly, well, pops up and grabs your attention. And by 'grabs your attention', I mean 'annoys you'. You can get rid of it by clicking on the 'close' or 'cancel' button in the pop-up's panel (or 'window'). Some crafty advertisers, however, create fake buttons that look similar; if you click on these, you're taken to that advertiser's **website**. In some cases, what you think is a genuine advertisement isn't; it's an attempt by **hackers** to infiltrate your computer. In this case, clicking on the 'close' button will actually

download a **virus** or **software** to gain unauthorised access to your computer.

Knowing how irritating (and possibly damaging) these can be, most internet **browsers** allow users to block pop-ups almost completely.

See also: **ad blocker**

Computer genius Ethan Zuckerman is considered the inventor of the pop-up ad. He later apologised for the unforeseen nuisance into which pop-ups evolved.

Post

(noun) The name for any message on **social media**. This could be text, a photo, a **GIF**, video or a **hyperlink**. For example, updating your status on Facebook is known as a post.

(verb) The process of creating a message on social media, a blog or forum.

See also: **share**

Profile

Think of this as your biography that appears on **social media**, **forums**, **chat rooms** and **newsgroups**; a means for people to identify you and get to know who you are. Profiles vary in both the information that's required and which elements of that information are displayed publicly, but usually include the following:

* Your name
* Username (the name by which you want to be known; this can be your real name but more often than not it's a nickname of your choosing)
* Date of birth (again, this can be your real DoB or your 'showbiz' age)
* **Profile picture** (this could be an actual photograph, another image or an **avatar**)

* Your location (this can be as vague or as precise as you want . . . but it's best not to use your actual address)
* Short biography (this can be serious or silly)

NB To enable people to find you easily it's best to use the same username and profile picture across all the social media, forums, etc., with which you engage.

Profile picture

This is the image you use on **social media**, **forums**, **chat rooms**, **newsgroups** – anywhere you interact with others **online** and need an image to represent who you are. Most people agonise over the photo they use since this tiny image is what others will judge them on. What starts off as an ambition to find a photo that conveys your beauty, intelligence, extrovert nature, humility or fun side usually ends up just with you getting frustrated, giving up and

uploading a badly taken **selfie**. Of course, your profile picture doesn't have to be a photo of you now; it can be a photo of you as a child or a baby. Or it can be a photo of your own baby, a beautiful sunset, your car, your cat, your favourite fictional character, your favourite food . . . anything that you feel represents your character or personality, as long as it's not obscene.
See also: **avatar**, **profile**

Program

In computing terms this is a structured set of instructions to a computer that tell it to perform a particular task. In essence, computers need a program to enable them to become more than just expensive paperweights. Programs can be games, word processors, **operating systems**, spreadsheets, calendars, web **browsers**, **email** applications . . . you get the idea. All **software** is a set of computer programs.

Q
QR code

QR code

This stands for Quick Response Code, a weird-looking barcode you've probably seen printed in an ad or on a product. It's designed to allow anyone with a **mobile device** to easily access a company's **website** and find out more information about a product or service without having to type in the whole **web address**. The 'quick' part of the name is questionable since to get the QR code to work you need to have **downloaded** a compatible QR reader **app** (of which there are many types) and then scan it with your device's camera. QR codes never achieved widespread popularity and are fast becoming obsolete, being replaced by **augmented reality** where images can be recognised automatically by your device's camera, directing the user to a website without the need to download any other app.

Some cemeteries use QR codes on their headstones to create what they call 'living memorials'. Scanning these will show the user a biography or tributes to the deceased.

From RAM *to* Rickrolling

RAM

When you're buying a computer or **mobile device** you'll often see these letters in the technical description. They stand for Random Access Memory, a piece of **hardware** that processes instructions and temporarily stores data from **programs** when your device is in use. RAM is measured in megabytes, abbreviated to MB – or gigabytes, abbreviated to GB (1 gigabyte is the equivalent of 1000 megabytes). In order to avoid getting too technical about RAM – and by 'getting too technical' I mean 'getting technical at all' – just remember that RAM is like money. It's better to have more than less.

Reboot

Sometimes your computer or **mobile device** might freeze or become unresponsive; no matter what you do, you can't get it to function properly. In cases such as this, think of a reboot as stage 3 where stage 1 is 'loud swearing', stage 2 is 'rising frustration' and a possible stage 4 involves a hammer.

Reboot sounds complicated but it's not; it's just the technical term for turning your computer or device off and on again. Rebooting allows the **operating system** and **programs** to reload, which usually – but not always – cures the problem.

Reddit

It's been said that there's perhaps no other **website** that's more informative and entertaining than Reddit. Think of it as an incredibly massive **forum** that's split into hundreds of thousands of topics called 'subreddits' that cover the most bizarre subjects imaginable such as birds **Photoshopped** so they have human arms, photos of things that are not interesting, people wearing horse masks, pointless stories, people who drink wine while showering, sexy images of toasters or photos of Slavs squatting.

You can find out almost everything here. There's a voting system where the best **content** moves up the Reddit rankings so that more people are able to see it and the unpopular or less interesting content disappears. The site has been called 'the front page of the internet' for good reason; nearly everything that's become big news or has gone **viral** is likely to have started here.

Visit: www.reddit.com

Retro

Anything that deliberately imitates a style or fashion from the recent past is retro . . . and by recent past we're usually talking about the 1930s onwards. It's a cover-all term that's overused to describe everything from clothes, hairstyles, packaging and graphics to furniture and furnishings, architecture, industrial design and even cars. The current MINI, Fiat 500 and VW Beetle are all considered retro designs that owe their styling to their predecessors from the 1960s (1940s in the case of the Beetle).

Designers through the ages have always drawn on historic influences but the concept of 'retro' as we know it today can also have the added element of kitsch, i.e. a degree of cheesiness or tackiness that can still be appreciated in an ironic and humorous way – just like the illustrations in this book.
See also: **vintage**

Rickrolling

This is pranking someone by tricking them into clicking on a **website** link that they think will take them to a topic of interest, but which in fact takes them to a clip of Rick Astley singing his 1987 hit 'Never Gonna Give You Up'. If you're a victim of this prank you're said to have been 'Rickrolled'. Although Rickrolling still happens, it reached a peak of popularity between 2007 and 2009. On 1 April 2008, **YouTube** got in on the joke by making every video link on its **homepage** into a Rickroll. It's been reported that Astley found the phenomenon 'bizarre and funny'.

Rickrolling is considered an internet **meme**.

From Safari *to* Swipe left/right

Safari

A web **browser** developed by **Apple** and which is the default browser on all Apple devices. There used to be a **Windows** version, but this was discontinued in 2012.

Satnav

If video killed the radio star then satnav killed paper maps. This is the name given to any system that uses **GPS** to locate your position on earth, and therefore enables you to find the quickest/shortest/most economic route to a given destination. Satnavs can be standalone devices for your car or an **app** that works on your **smartphone**, for example, **Google** Maps. Most satnavs will work out alternative routes if there's a problem on the one you're taking while some also provide traffic warnings and car-park availability in real-time.

SCART lead

A SCART lead or connector is a chunky French industry-standard connector that dates back to 1977 and which usually linked your TV to a video recorder, a games console or, later on, a DVD player. I say 'linked' but SCART leads were not the most secure of connections, often becoming dislodged just as soon as you even looked at them. They have mainly been replaced by **HDMI** cables.

Screengrab

Say there's something on your **mobile device** or computer, of which you need a visual record. It could be an interesting feature on a web page you want to reference later, or it might be an amusing exchange of **text messages** you want to **share** with friends. Depending on the device you're using, there'll be a simple way to take a photo of what you see on the screen. This image is then known as a screen-grab or screenshot, and can be shared or stored as you would an ordinary photo.

Screensaver

A computer **program** that blanks the screen or fills it with moving images after a period of inactivity (screensavers are built-in to your **operating system** or you can usually choose your own image). The original purpose of the screensaver was to prevent a static image from burning itself into the old style cathode-ray tube computer monitors. Modern screens don't have this problem but screensavers are still used mainly for decorative or entertainment purposes. After all, who wouldn't find whimsy in the idea of their resting screen resembling a tropical fish tank?

Screen time

This is the amount of time spent using a device such as a **smartphone**, computer, video game console or a TV. Too much screen time is generally regarded as being bad for children's health, although there aren't enough in-depth studies to come up with definitive results and recommendations. Effects linked to spending too much screen time include sleep disturbances and an increase in weight. One US study found that for every hour spent in front of a TV, children consume an extra 167 calories.

Search engine

You're about to read the understatement of the book: there's a lot of information on the internet.

It's estimated there are over 1.5 billion **websites** and 130 trillion web pages so trying to find the exact information you need can be quite challenging. That's where a search engine comes in. These are special websites that have scanned and indexed all these other websites so you don't have to. To find the information you need, enter the keywords or phrase into the search box (e.g. 'what is a search engine?', 'handsome authors named Mark') and press 'enter' or 'search'. The search engine will then provide a ranking of all the websites that contain information that meet your criteria.

Most readers will be aware that the most popular search engine is **Google**, which has almost 91 per cent of worldwide market share across all **devices**. Its nearest rival on desktop computers and **laptops** isn't that well known in the West; it's Baidu, known as 'the Google of China', with a market share for these devices of about 14 per cent. Other popular search engines are **Bing**, **Safari** and **Yahoo!**

Selfie

This is a self-portrait, usually taken at arm's length with a **smartphone**, and usually **shared** over **social media** sites. The term applies to a shot of one person as well as a shot of multiple people – as long as it's taken by someone in the group.

The concept of selfies really became popular after the introduction of the first phone with a front-facing camera in 2003. Their appeal comes from how easy they are to create and share, and the degree of control they give the photographer as to how the image is finally presented (i.e. cropping it or using **filters** in order to create the most flattering shot).

See also: **selfie stick**

Ten things you may not know about selfies

1 In 2012 *Time* magazine considered 'selfie' as one of the 'Top 10 Buzzwords' of that year. In 2013 'selfie' was announced as being the 'word of the year' by the *Oxford English Dictionary*.

2 In 2014 'selfie' was officially accepted as a valid word in Scrabble.

3 A poll by Samsung in 2013 found that selfies account for 30 per cent of all photos taken by eighteen to twenty-four year olds.

4 The pop-up Museum of Selfies in Glendale, California, opened in May 2018; a place where people could 'explore the history and cultural phenomenon of the selfie'.

5 A selfie taken by host Ellen DeGeneres during the 86th Academy Awards featured her with twelve other Oscar celebrities. By the end of the ceremony it had been **retweeted** over 2 million times.

6 In 2015 it was reported that more people had been killed taking selfies in dangerous situations than had been killed by sharks.

7 Yellowstone National Park in the US issued a warning about taking selfies with bison after five visitors were gored.

8 The 'selfie effect' refers to the distortion of the nose in close-up photographs, caused by the nose being closer to the camera than the rest of the face.

9 'Selfitis', the obsessive need to **post** selfies, has been recognised as a genuine mental disorder. This follows extensive research by Nottingham Trent University's Psychology Department.

10 A 2015 survey reported that people born after 1980 will take an estimated 25,000 selfies in their lifetimes.

Selfie stick

A telescopic pole with a holder for a **smartphone** at one end and a trigger mechanism at the other so you can take a selfie at distances and angles to which your arms won't stretch/contort. The **selfie** stick achieved mass popularity in 2014, the same year it received two quite different accolades: *Time* magazine listed it in its round-up of the year's best inventions while the *New York Post* named it as the year's most controversial gift (commentators criticised the selfie stick as an aid for the vain, nicknaming it the 'Wand of Narcissus').

The selfie stick has been banned in a number of public places for reasons of safety, nuisance value – and because they can be used as weapons.

Server

Unlike the computer you use on an everyday basis to write documents, order shopping, play a word game that looks remarkably like Scrabble, check how many people liked the photo of the sunset you took in Lanzarote, etc., etc., a server is a computer whose whole raison d'être is to handle specialised tasks and deliver data to other computers – usually over a **network** or the internet. Examples include web servers, which deliver web pages, and mail servers, which facilitate the sending and receiving of **email** messages.

Set-top box

You'll probably know it as your Sky Q, **Apple TV**, Freeview or Virgin **TiVo**, but 'set-top box' is a generic name for any device that allows your TV set to receive digital TV signals from your cable or satellite TV provider. Named after a time when TV sets were chunky enough to allow the box to rest securely on top, most of these devices feature a **DVR**. Some also respond to voice commands.

Sex tape

Not a movie like *Buffy the Vampire Layer* or *Forest Hump*, but the term given to a video made of a sex act usually (but not exclusively) involving at least one C- or D-list celebrity. These tapes have a habit of falling into the hands of the media or turning up on **social media**.

Celebrities on sex tapes include Pamela Anderson (twice), Paris Hilton, Kim **Kardashian**, Colin Farrell, Katie Price, Rob Lowe, Usher and Hulk Hogan.

Sexting

The process of sending someone messages with sexual content or sexually explicit photos/videos of yourself. Despite the name, sexting doesn't have to take place via a **text**; it could be via any messaging **app** or **online**. Sexting might seem like a good idea at the time, or even a laugh, but these intimate messages or images have a habit of being discovered some time later by someone you really don't want to see them, i.e. your partner. This person will invariably see a dubious photograph on your device and ask about the sender (and by 'ask' I mean 'shout'), 'Who the hell is Alice?!!!' (NB Insert any name here.)

Share

Used as a verb or a noun, this refers to sharing a **post** or material on **social media** (it could be from an individual or a company/organisation). It can refer to your own **content** that you have shared, but often refers to someone else's post. What you share could be someone's personal views or observations, information, a news article, an advertisement, a photo, a video, a **hyperlink** – anything that someone else has previously posted. But why share the post in the first place? It can be because you find the content entertaining or useful, because it reflects your own views or supports a cause you believe in, or it might be a way of self-expression. Sharing has been called the 'currency of social media', and just like real currency, the more you share – or the quicker you share – the more popular you can become.

See also: **viral**, **viral marketing**

The various social media **platforms** have different names for sharing, e.g. **Facebook** calls it 'share', **Instagram** calls it 'share' or 'regram', while on **Twitter** it's a 'retweet'.

Shazam

You hear a song on the radio and you're racking your brain trying to work out what it is and who it's by. That guitar riff is familiar. And so's that haunting vocal. You've got it! It's Fleetwood Mac. No it's not. It's the Vengaboys. No. It's definitely Metallica. But then again . . .

Shazam (now owned by **Apple**) takes the guesswork and frustration out of this type of situation. It's a free mobile **app** that recognises pre-recorded music (so nothing played live, and not you humming or singing no matter how good you are).

How does it work? Well, it uses the built-in microphone

in your computer or **mobile device** to take a ten-second sample of that song. From this it then creates what's called an acoustic fingerprint, which is compared instantly to over 11 million songs in its **database**. If it finds a match, it sends information such as the artist, the song title and the name of the album back to the user.

Five things you may not know about Shazam

1 Although there are rivals to Shazam, it's the leading music recognition app. It's been **downloaded** more than 1 billion times and used on more than 500 million mobile devices.

2 The most Shazamed track of all time is 'Wake Me Up' by Avicii (Shazamed over 23 million times).

3 It started off as a call-in service in the UK. You dialled 2580, let Shazam listen to the song and then hung up. You'd then get a **text message** with the song details.

4 In 2008, it became one of the first apps on Apple's **App Store** (it's now built into **Siri**).

5 It sometimes struggles with giving exact details for classical music as the same piece can be recorded hundreds of times by different artists/orchestras.

Silent disco

Imagine going into a room full of people dancing to nothing. If your first instinct is that you've ventured into a Cheeky Girls/Jedward gig you'd be close, but wrong. What you're observing is a silent disco: a disco where the music is played over wireless headphones worn by the participants, rather than through loudspeakers. Silent discos mean that it doesn't matter where you are at the venue, everyone hears the sound the same crisp and clear way. In some cases, the DJ will use different channels to play different music, which means that members of the audience can be dancing to different tracks at the same time.

Why silent discos? Well, it's a way for concert organisers to get around any noise curfew issues, plus it makes a music event, well, more of an event (albeit an event where people look silly).

SIM card

This is the small, flat piece of metallic plastic about the size of a finger nail that's inserted into the side of your mobile phone (or behind the battery), which identifies you to your mobile network provider, allowing you to make use of that network e.g. to make calls and send data. When you get a new phone you'll either be given a SIM card to insert or asked to transfer the one from your old phone as long as it physically fits (the older the phone, the larger the SIM card; original SIM cards were 85mm x 35mm. Current Nano SIMs are 12.3mm x 8.8mm). Apart from enabling your phone to function, the SIM card can also store contacts and messages. **Tablets** that connect to a mobile phone network also use SIM cards in the same way.

You don't need to know what SIM stands for (and hardly anyone does) but in case you want to impress your friends or win a pub quiz, it's Subscriber Identity Module.

Simulcast

This is a live TV or radio programme that is broadcast simultaneously over a conventional broadcast channel and the internet.
See also: **live streaming**, **streaming**

Siri

Built into its **operating system** and responding to the **wake word** 'Hey Siri', this is **Apple**'s **digital assistant**, the equivalent of **Amazon**'s **Alexa**. As you'd expect, Siri is slightly biased towards Apple products. Ask her 'What phone is best?' and you'll get responses such as, 'Wait . . . there are other phones?' or 'Seriously?'

Site

See: **website**

16:9 aspect ratio

TVs have different screen aspect ratios (or formats) that indicate the proportion of the picture's width to its height. Back in the day, the old boxy TVs had an aspect ratio of 4:3, i.e. almost square. Nowadays, nearly all modern TVs are 16:9 (also called widescreen), which is good – since all TV programmes are made and delivered in this aspect ratio.

Every wondered why the screen grows wider at the beginning of a movie? It's because the commercials shown beforehand follow the TV aspect ratio of 16:9 while the movie itself was probably filmed in the cinema format of 21:9.

Skype

You probably already resent the fact that your children emigrated to Melbourne or California knowing that it's too far for you to visit them, and then there's the extortionate cost of international calls . . . Well, Skype enables you to call them as often as you want. Not only can you take pleasure in knowing that this will annoy them, but the other advantage is that if they're also a Skype user, it's completely free!

Put simply, Skype is one of the best-known and most popular internet video-messaging services, a way to make and receive free voice or video calls **online**. Just **download** the Skype **app** on to your computer or **mobile device** and you're all set to ask, 'What's the weather like?' and 'Why don't you call me?'

Visit: www.skype.com

See also: **video conference**

Alternatives to Skype include **Apple**'s **FaceTime** and **Google Duo**.

Smart home

That's smart as in intelligent, and by intelligent we mean a home that features a wide range of automated systems to make life easier. For example, opening and closing the blinds depending on how sunny it is, learning when you leave and enter the house each day and adjusting the temperature accordingly, feeding your dog or cat at set times, vacuuming the house after you leave for work, turning off the alarm, switching on the oven and unlocking the front door as soon as your car pulls into the drive, watering the garden when the soil becomes too dry, cutting the lawn when the grass becomes too high, ordering more milk before you run out, dimming the lighting when you play 'Let's Get It On' by Marvin Gaye . . .

These are all things a smart home can do (and if not now, then very soon).

It does it by a series of **WiFi** enabled devices that are connected over the internet and which communicate with each other, and the householder, via a main controller called a **home hub**.

Control of the various systems is carried out via a smart-home **app** or a **digital assistant**.
See also: **domotics**, **Internet of Things**

Smartphone

Introduced in the early 2000s but popularised by **Apple**'s **iPhone** when it was launched in 2007, these are mobile phones that function similarly to conventional computers, i.e. they have full colour **touchscreens**, can store music, photographs, videos and documents, access **email** and the web via a mobile phone network or **WiFi**, and also run various **programs** (called **apps**).

Oh yes. You can also make and receive phone calls and text messages.

Nowadays, most mobile phones will have some degree of smartphone capability.

Smart TV

Not TVs that look chic and classy, smart TVs are those with internet connectivity, which means you can access **video on demand content** from broadcasters such as BBC **iPlayer** or 4OD, as well as **streaming** services such as **Netflix** or **Amazon** Prime Video – as well as the ability to connect to other devices such as **smartphones** or **tablets** and, of course, browse the web. To make this happen, your TV will need to be within range of a **WiFi** network or, if you want to ensure a stronger, more stable internet connection, you can plug your TV directly into your **broadband router** via an **ethernet cable**.

Most TVs on sale these days are smart.

See also: **streaming stick**

Smartwatch

Think of this less like a watch and more of a mini computer worn on the wrist. Sharing a lot of their technology with **smartphones**, smartwatches have **touchscreens** and are supported by a wide range of **apps**. So what can they do that other watches can't? The main thing is that they can display notifications from your smartphone (e.g. forthcoming appointments, **emails**, **texts** and **social media** updates), you can make and answer calls on them and, as mentioned previously, access a huge number of apps. Many smartwatches also include a heart rate monitor and a pedometer so you can track your workouts. They also allow you to change and customise the watch face and, lastly, **Apple**'s own smartwatch (called, unsurprisingly, the Apple Watch) allows you to pay for things using **Apple Pay**.

SMS

Another name for the 'standard' **text message** that all mobile phones can send. It stands for Short Message Service, so called because SMS messages are limited

to 160 characters (longer messages are split up into multiple messages) and can only contain text.
See also: **MMS**

Snapchat

This is both a **social media site** and an **instant messaging app** – but it's the messaging aspect that makes it unique. In common with other social media, you connect to friends but rather than using it to send messages (although you can of course do that), Snapchat is primarily used to send photos and short videos, up to ten seconds long. What makes it unique is that, once opened, the image disappears after a few seconds. Forever.

Why? Well Snapchat is designed to communicate 'in the moment' – hence the image vanishes almost as soon as it appears.

The advantage of this 'self-destruct' nature is that because the photos or videos are sent straight from your **smartphone** there's less anxiety about worrying about how you look and, also, how many people will like it. The disadvantage is that it can lead some people to be a bit more daring; they can use Snapchat for **sexting**, in the knowledge that if they send someone a provocative photo of say, their private parts, there's no lasting evidence.

Except there is.

The image or video disappears, but if you're quick enough you can take a **screengrab** of it – although Snapchat will notify the person sending the image that this has been done.

Five things you may not know about Snapchat

1 The idea for the disappearing images came to Snapchat's founders when a friend of theirs regretted sending a photo to someone else.

2 When Snapchat launched in 2011 it was called 'Picaboo' (it was rebranded as Snapchat in 2012).

3 The most common images snapped are drinks (a survey found that 93 per cent of users have sent a photo of their drink to someone else).
4 It's been reported that 3 billion photos are snapped each day.
5 Seventy per cent of users are female.

Social media

Like **fake news** and **cryptocurrency**, the term social media is heard daily – and usually several times a day. There seems to be as many people trying to define what it is as there are using it. So here goes . . .

In essence, the term refers to a collection of free to use **websites** or **apps** for people to **share** information and interact with each other (i.e. the 'social' aspect of the term). In this way, each social media **platform** is an **online** community consisting of two types of people: those who are known to you (family, friends, colleagues) and those who you don't know at all. This latter group might be friends or associates of the first group – or they might be complete strangers. It's this aspect that's a blessing as well as a curse. Yes, you can keep in touch almost instantly with people you know but, equally, it gives random strangers the ability to act in undesirable ways, for example, by **posting** inappropriate messages, **cyberbullying** or **trolling**.

The information you share on social media might be a short message, longer prose, a video, a photograph, a news story, music, a website, a presentation or an **animated GIF**. Or it might be a **hyperlink** to any one of these. Some social media sites are geared towards specific uses or audiences; for example, **LinkedIn** is aimed at business professionals, while **Instagram** focuses on sharing photographs or videos.

Regardless of which type of social media you're using, they operate in more or less the same way. You register your details and set up a **profile** (some of this information will be visible to others). When you're set up, you search for people that you know in order that you can connect/interact with them. Then you post your message, photo, video, etc., and wait for people to comment on it – or you comment on someone else's message, photo, video, etc.

There's so much to say about this subject that the topic could easily continue for several more pages.

Don't worry, it's not.

All that's left to add is that what seems like quite a trivial, almost hollow and meaningless way to spend your time (getting upset in the process when only five people have 'liked' the photo of your cat wearing a fez), social media isn't just big business, it's massively, immensely, colossally huge. The best-known social media site is **Facebook**, which has more monthly users than the entire population of China and the US combined, and which employs over 25,000 people worldwide and enjoyed revenues in 2017 of over $40 billion. In 2018 (depending on what report you read) between 72 and 79 per cent of the world's online population uses a social media platform.
See also: **social networking**

Monthly average users for popular social media sites

Facebook	2234 million
YouTube	1500 million
WhatsApp	1500 million
Instagram	813 million
Tumblr	794 million
Reddit	330 million
Twitter	330 million
LinkedIn	260 million
Snapchat	255 million

Social network

Another name for **social media**, but it's slightly different from the concept of **social networking**.

Social networking

This is where it gets confusing. Some people claim that social networking is the same as **social media** and use both terms interchangeably. Others harp on about how they are in fact two different concepts, albeit with some crossover. The whole issue is made cloudier since a social media site can also be called a **social network**...

Rather than individuals, it's businesses (and marketing people who want to appear too cool for school) that tend to distinguish the terms far more than the rest of us, so with the intention of keeping it simple, think of social networking as the principle of engaging with an audience, usually building a two-way relationship through things such as **blogs**, **forums** or **webinars**. Social media on the other hand is a tool to communicate your message to a mass market (i.e. everyone views the same **content**), and where generating 'buzz' and sharing it is an important objective.

And that's one way to differentiate them.

If you want to.

Software

This is a collection of **programs** that 'tell' a computer how to perform a certain task. In basic terms the software provides the instructions for what to do and how to do it. An **operating system** such as **Android** or **Windows** is software, as are specific programs such as **Word**, Excel, PowerPoint or **Photoshop**.

See also: **hardware**

Spam

Another term for **junk mail**, the term spam is a tribute to the 1970 *Monty Python* sketch involving a café where every single dish includes spam (the canned meat). Spam is mentioned over 130 times; symbolic of the unrelenting nature of the electronic variety.

Spotify

One of the most popular **music streaming** services, which provides access to over 35 million songs. The service is available free (with advertising between tracks), or via a monthly subscription (no ads, more songs, better quality and the ability to **download** songs so you can listen to them **offline**).

Spyware

One of the sneakiest types of **malware**, this is the generic name for **software programs** designed to lurk undetected on your computer and secretly monitor activities with the aim of capturing data such as usernames and passwords or **online** bank account details and credit card information.

Sometimes the only sign that your computer's been infected with spyware is when it slows down; that's because while it's sending the stolen data to the criminals behind it, the spyware is reducing processing power or internet speed. *See also:* **cybercrime**

SSD drive

SSDs (solid state drives) perform the same function as a **hard drive** but the data is stored on what's known as flash memory chips, which have no moving parts. They work on the same principle as **flash drives**, but they're a lot faster and have much more capacity. Some computers give you the option of specifying whether you want it to have a hard drive or an SSD – the difference being capacity and cost (SSDs are more expensive).

Steampunk

What started as a science-fiction genre that fused the functionality of modern technology with a Victorian look and feel (i.e. industrial and elaborate) has evolved into a whole **retro** design philosophy and visual style. This involves taking new technology then combining it with traditional Victorian materials and a dash of whimsy. A computer keyboard that uses old typewriter keys mounted on a brass and mahogany casing? That's steampunk.

Sticky content

Unlike the real world, where anything sticky isn't a good thing (apart from toffee pudding), in the **online** world, it's very desirable. Used in relation to **websites**, sticky **content** is anything that attracts visitors, holds their attention, gets them to spend more time there – or encourages them to return. To do this, the content needs to be relevant, engaging, relatable, newsworthy and credible. Video content, **vlogs** and **blogs**, for example, make for good sticky content.

Streaming

There are two ways to watch a video or listen to music over the internet. One is to **download** it on to your computer/ **mobile device**, which means that you can't access it until the whole **file** is downloaded. Streaming, or internet streaming, on the other hand, is a technology that allows video and music to be delivered over the internet as a continuous flow. That means you can watch or listen almost immediately. Companies such as **Netflix** and **Spotify** deliver their **content** by streaming; it's quite a recent development, made possible by **broadband**, which allows for faster and better quality internet connections.

The other main difference between streaming and downloads is that, after a download, the video or song is saved to your computer and will remain there until you choose to delete it; with streaming, the data is deleted after you use it.
See also: **live streaming**, **simulcast**, **streaming stick**, **video on demand**

Streaming stick

A small device, a bit larger than a **flash drive**, that you plug into an **HDMI** port at the back of your TV to turn it into a **smart TV**. Popular makes of streaming sticks are the Roku, **Google**'s Chromecast and **Amazon**'s **Fire TV Stick**. There are no monthly fees for these devices; you pay for **content** from the content providers where relevant.
See also: **streaming**

Surf the net

What sounds more dynamic? 'I'm going **online** shopping for a new cardigan and then visiting the caravanning club **forum**' or 'I'm surfing the net!'?

Er . . . it's the second option.

Surfing the net is the sexy-sounding way of describing spending time on the internet, whether it's checking **emails**, using **social media** or looking at **websites**.

Swipe left/right

This refers to the way you react to possible matches on the **online dating app Tinder** (and some others). Swiping right on the screen when you're presented with a potential date's photo/biography indicates you like what you see. Swiping left means thanks but no thanks.
See also: **catfish**

From Tablet *to* Twitter

Tablet

Also known as 'tablet computers', this is the generic name for a slim, portable computer with a **touchscreen** and rechargeable battery (some also use a stylus as well as a keypad), e.g. Samsung's Galaxy Tab, **Microsoft**'s Surface and **Apple**'s **iPad**. Tablets have the same functionality as a **smartphone** but with larger screens (usually seven inches or more measured diagonally). Many tablets can be used with separate keyboards connected via **Bluetooth**.

Tablets were portrayed in numerous works of science fiction including *Star Trek* (1966) and *2001: A Space Odyssey* (1968), more than thirty years before they first appeared.

Tag

Tags are popular in **blogs** and on **social media** as a way to identify and engage with a person, a business or a group – or alert others to **content** they might find interesting. Think of a tag as a sort of **hyperlink**.

Blogs: Say you're **posting** a recipe for your legendary prune and fig pie on your food blog. To enable people to find it easily – and find articles on your blog of a similar nature – you might use the tags: 'prune', 'fig', 'pie', 'dessert' and 'laxative'.

Social media: You can tag people to identify them in a photo, for example, or alert them that there's a post relevant to them (or sometimes about them). This is usually done by inserting an '@' symbol before their name. You can also tag 'things' (e.g. places, concepts, objects) so that, similar to blogs, it's easy for users to find content on the same subject. Although each social media site has its own way of tagging, they all follow the same general principle.

See also: **hashtag**

#TBT

Throw Back Thursday; an opportunity (or excuse) to use that **hashtag** to **post** nostalgic 'Back in the Day' photos on **Twitter** or **Instagram**. An old car, old friends or colleagues, ridiculous fashions, even more ridiculous hairstyles – the funnier (and by funnier, I mean cringeworthy) the better.

Tesla

The company funded by Elon **Musk** but also the name of his range of electric vehicles. Yes, there are lots of electric cars out there but there's a mystique about Tesla cars that derives from a combination of their cost (high) and their technology (also high), and also because they're built by a technology company rather than a conventional car manufacturer. There's also the Tesla philosophy that the cars can be environmentally friendly yet deliver high performance. Its sportiest Model S has a mode called 'Ludicrous+'; tested by *Motor Trend* magazine, it accelerated from 0 to 60 mph in under 2.3 seconds (faster than a Porsche 911 Turbo). Tesla is also heavily involved in the development of **autonomous cars**.

Tethering

You want to connect your computer (usually a **laptop**) or **tablet** to the internet but there's no **WiFi**.

No problem! As long as you have a mobile phone and a strong mobile signal (preferably **4G**) you can use it to provide internet access to the other device. This is called tethering.

Tethering, as the name suggests, can be done by connecting your phone to the device by a **USB** cable; however, it can also be done wirelessly via **Bluetooth** or WiFi. Whatever method you use, you first need to switch on 'Personal hotspot' or 'Portable WiFi hotspot' on your phone (different phone manufacturers call it different things). You then need to connect your phone to the device via Bluetooth, a USB cable or WiFi (in this case, select your phone as the WiFi network).

It all sounds a bit complicated but it's not really.

Texting/text messaging

Yes, I'm sure you know exactly what this is but in case any readers don't, it's sending messages between two or more users usually on phones or **tablets** – but it's also possible on computers and **laptops**. Some text messages allow you to also include images, videos and sound.

See also: **MMS**, **SMS**

Thread

Used in conjunction with **emails**, a thread is a running list of the back-and-forth conversations in that same email. Arranged in chronological order so the most recent response is usually at the top, the thread provides an easy way to keep track of who's said what. The term is often used in the following way:

Friend: 'Where are we meeting?'

You (exasperated): 'It's the first thing I typed when we began emailing about arrangements. Check the thread!!!'

3D printing

Seeing this in progress is really like something out of science fiction – the ability to create solid objects from a digital **file**. Instead of ink, a 3D printer controlled from a computer uses a range of materials to create the three-dimensional object. In most cases it's plastics and resins but they can also use metallic and ceramic powders. The process works by turning the object you want to create into thousands of very thin slices. Like an ink-jet printer that goes back and forth across a page building up an image or a document, the 3D printer goes back and forth to gradually build up physical layers. It's the combination of each of these tiny slices one on top of another that makes the finished object.

Uses at the moment include rapidly produced prototypes of products and making everything from architectural models, footwear and custom jewellery to automotive and aviation parts. In the medical world 3D printers can be used to create teeth, custom implants and prosthetics, and also reconstruct bone. There are even 3D printers that use edible ingredients to print food.

TIFF

See: **JPEG**

Tinder

An **online dating app** that's remarkably simple and easy to use but, because of this, has gained a reputation for being addictive, attracting superficial users who are searching for a free hotel room more than a relationship, and encouraging casual sex.

After setting your search criteria you're presented with a selection of possible dates that are located near you. If you like someone you see, you swipe their photo right. If you don't, you swipe it left (which is the equivalent of saying, 'It's not you it's me,' without the awkwardness).

If your would-be date also swipes you right, you're notified – and you can get in touch.

The top five 'swipe-right' jobs on Tinder (as of March 2018)

	Men	*Women*
1	Pilot	Physical therapist
2	Entrepreneur	Interior designer
3	Firefighter	Entrepreneur
4	Doctor	PR/Communications
5	Media personality	Teacher

TiVo

A generic name for **DVR**s, named after the first consumer DVR that was launched in 1999 by the TiVo Corporation. Its popularity has caused the brand name to also be used as a verb (e.g. 'Don't panic! I've TiVo'd *The Jeremy Kyle Show*. We can watch it later').

Touchpad

Laptops are designed to be compact and used in small spaces – such as on laps and the tables in coffee shops, for example. Using a mouse to move a cursor round a screen can therefore be difficult, especially if there's no suitable surface for it. That's why all laptops sold today feature a touchpad (sometimes called a trackpad); a small area below the keyboard that contains touch-sensitive areas that respond to finger movements – translating these movements to the cursor on the screen. Multitouch touchpads can sense multiple fingers so different movements (e.g. pinching, tapping, dragging and holding) will have different effects.

Touchpads on laptops replaced the trackball. This was a pointing device that consisted of a small ball that protruded from the palm rest, and which was moved with your fingers, while you simultaneously clicked on mouse buttons nearby. It was actually much more intuitive and simple to use than it sounds.

Touchscreen

Looking like a regular TV or computer screen, a touchscreen is a way to interact with a device without using a keyboard or mouse. You'll find them on **smartphones**, **smartwatches** and **tablets** and things such as **satnavs**, information kiosks, ATMs, self-service checkouts, etc.

Trackpad

See: **touchpad**

Trending

Before **social media** made the reporting of news almost instantaneous, things that everyone suddenly started to talk about might have been referred to as being 'popular' or 'hot news'. Now they're referred to as 'trending' on social media. Trending can refer to the most popular topics of conversation anywhere in the world: the arrest of a celebrity, a natural disaster, an act of terrorism, an act of heroism, an outspoken remark from a president, an unexpected sports result, the finale of a long-running TV series – all of these can be (and have been) top trending stories.

TripAdvisor

An international travel and restaurant information and review **website**, and the largest travel site in the world with 315 million users. Like **Trustpilot**, TripAdvisor is open to fraudulent reviews from users and the businesses listed; however, it tries to authenticate each review before

it's **posted** by checking the user's **email** and **IP address**. If it feels a business is posting suspicious reviews, it can blacklist it from the site. (In 2011 it discovered that a Cornwall hotel had offered guests incentives to post favourable reviews.)

Five of the oddest/funniest review titles posted on TripAdvisor

1 'Stay out – it's haunted'
2 'The owner kept trying to sell me the hotel'
3 'We got spewed on'
4 'Dead cockroach floating in my drink'
5 'Poo in the kettle'

Troll

Also known as internet trolls, these are obnoxious and malicious internet users who deliberately stir up trouble and distress others (usually people they don't know) by **posting** inflammatory and often hurtful posts on **social media**, **forums**, **chat rooms** and **newsgroups** – basically anywhere people are able to post comments. The action of doing this is known as 'trolling'.

Why do they do this? Well, firstly, trolls take great pleasure from seeing people's outrage or causing anguish and, secondly, because they can. Examples include posting hurtful messages on tribute pages set up after someone's death or threatening people over their political or religious views.

Like the mythological troll, the ugly, angry and disruptive creature that conceals itself in dark places, trolls hide behind the anonymity of the internet and it's often very difficult to identify them. Trolls thrive on getting a reaction to their comments so the best way of dealing with them is by ignoring them.

See also: **doxing**, **flaming**, **hater**

Trustpilot

A **website** that publishes reviews for **online** businesses. Anyone can **post** a review (you just need to register with an **email** address or **Facebook** account); about 500,000 new reviews are posted each month. A criticism of the site is that it's easy for employees to post fake good reviews for their own business, and bad ones for rivals. It therefore relies on users and the businesses themselves to identify and report any questionable reviews, which can then be deleted if they're felt to be fraudulent. (Businesses can publicly reply to any bad reviews with their own side of the story.)

See also: **net rep**

Tumblr

A popular **microblogging website** that tends to be used by those wanting to express their creativity or artistic side.

Visit: www.tumblr.com

Tweet

The name for a message or **post** on **Twitter**, so called because at only 280 characters (previously, the limit was 140 characters), it resembles the short chirp you'd hear from a bird. Reposting someone else's message on Twitter is known as 'retweeting'.

Apart from text, tweets can contain photos, **GIFs**, video and **hyperlinks**.

Twitter

One of the best-known **social media platforms** where people can **post** short messages of up to 280 characters, called **tweets**. Because the messages are so short they naturally need to be 'to the point' and their brevity means you can check what other Twitter users (individuals or organisations) are saying or doing with a brief glance; perfect in this attention-deficit world.

The way you see what other Twitter users are saying is by subscribing to their Twitter feed. Once you've done this you're known as one of their **followers**.

See also: **blue tick**, **hashtag**

Twitter accounts can be recognised by a username preceded by the @ symbol, e.g. @realDonaldTrump, @Lord_Sugar, @markleigh99.

From Uber *to* USB stick

Uber

You used to get a cab, now you take an Uber: the minicab service that's as convenient as it is controversial. Bookings are made via its **website** or mobile **app** and can also be scheduled in advance. You determine where you want to be picked up, where you're going and even the type of vehicle you want. You're also advised of the cost of the trip. All you have to do is wait, get in and then get out. The driver uses **GPS** to navigate to your destination and as soon as you've been dropped off, the payment is charged to your nominated bank card or **PayPal** account, and you receive the receipt via **email**. You can rate your driver afterwards and even tip him/her. One thing to note: the cost of the trip varies according to demand; it's what is called 'dynamic pricing'.

Uber operates in 633 cities but has been criticised by rival taxi companies who complain that it is allowed to bypass local safety and licensing laws (including satisfactory vetting of its drivers), which amounts to unfair competition. After various protests and legal challenges, Uber has been banned from operating in certain cities and even entire countries including Denmark, Bulgaria and Hungary.

Apart from regular cars, Uber has also offered rides in a DeLorean DMC-12 (the one from *Back to the Future*), a Lamborghini Gallardo and even helicopters. During National Ice Cream Month in the US, Uber users in certain cities can summon an ice-cream van with purchases billed to their account.

Unfriend

Although a term primarily used in relation to **Facebook**, unfriending someone means removing them from a list of 'friends' from any **social network**. The person you've unfriended won't be able to view any **posts** you've deemed private, but they will be able to see your **profile** and your public posts including photos. Unfriending someone can be done via the privacy setting on your account and is a less drastic way of distancing yourself from someone than **blocking** them. In most cases, the person you've unfriended won't be notified – but reciprocally, you'll be removed from their friends list as well.

If you change your mind (like you decide you really miss their posts that are inanely banal or perpetually smug), then you can easily add them as a friend again.

If only changing your friends was as simple in real life.

Upload

The opposite of **download**, this means sending **content** from your computer or **mobile device** to another computer or device, or a **website**.

URL

Unlike a **domain name**, which is the name of an overall **website**, a URL (Uniform Resource Locator) is a longer **web address** used to locate a specific page or section within that website. It's easy to understand if we take a company called, for example, Generic Company Inc. Its domain name might be 'genericcompany.com', while the section where you can find out about the history of the company can be found at 'www.genericcompany/history' or the section where you can **email** individual directors can be found at 'www.genericcompany/aboutus/directors/contact'.

To be honest, most people use the term URL when they actually mean domain name. It doesn't really matter.

See also: **IP address**

USB

Short for Universal Serial Bus, this is a current industry-standard type of connection that allows **peripherals** (e.g. **webcams**, **flash drives**, joysticks, scanners, printers, etc.) to connect to computers. Its other main use is as a means of charging portable devices.

Just when you thought it was simple there are different types of USB connectors and ports. Without getting too bogged down, there's the 'standard' USB port you'll find in computers and on flash drives, a smaller 'micro USB' port found in some mobile phones and cameras, and the new 'USB-C' port that's the emerging industry standard thanks to its much smaller size.

USB stick

See: **flash drive**

From Vaping *to* Vlog

Vaping

This sounds like a highly stylised 1980s dance move but it isn't; it's a term used to describe smoking **e-cigarettes** and also a range of similar devices designed to make smoking safer. These can resemble pens, asthma inhalers or what look like small wind instruments.

Vevo

If you view a lot of music videos on **YouTube** you may notice that you're often taken to that band or artist's Vevo channel. This is a **website** set up by the three largest music and entertainment companies (Universal, Sony and EMI) that acts as a video **hosting** service for their artists, paid for by advertising. The site doesn't just feature music videos, there are behind-the-scenes features, interviews and also original series. Vevo features over 50,000 videos from a whole host of bands you've probably heard of (e.g. the Beatles, Michael Jackson and Steps) to those you probably haven't (e.g. Chibbz, Mozzy and Moneybagg Yo).
Visit: www.vevo.com

Artists earn approximately a quarter of a cent each time their video is **streamed**, before the record company takes their cut. This might not seem like much but when you consider that 'Sorry' by Justin Bieber has been streamed 2.9 billion times, that's a lot of quarter cents. Over $7.25 million in fact . . .

Video conference

A meeting held between two or more people in separate locations, using the internet to provide a video link between those locations. So, instead of just talking to the other people in the meeting, you can see them as well (although that doesn't necessarily make the meeting any less tedious).

Skype can be used for group video conferencing.

Video on demand

Remember when we were slaves to broadcasters? They'd show our favourite TV shows and dictate on what day and at what time you had to watch them. Those bullying tyrants! Well, thanks to the advent of **streaming**, you can watch what you like, when you like – and don't let anyone tell you otherwise! Companies such as **Netflix**, **Amazon** and a host of cable/satellite TV operators all provide a video on demand service (also known as VoD).

Vimeo

A video-sharing **website** where users can **upload**, **share** or view videos much the same as **YouTube**. The main difference is that while YouTube has videos about almost everything, Vimeo is more discerning with a bias towards quality rather than quantity. It's regarded more as a professional network of contributors; many of its videos are from artists, musicians and indie filmmakers, and Vimeo staff curate their favourite videos.

Another difference is that YouTube is free for all users since the company earns its revenue from ads that you can either skip or sometimes have to endure before your chosen video plays. Vimeo has no ads, so earns its revenue by charging regular users to upload videos.

So, if you want to upload, share or view a video of your cocker spaniel walking upright on two legs and saying what sounds like, 'I love you', then use YouTube. If you want to **post**, share or view a video that could be considered artsy or informative, use Vimeo.

Visit: www.vimeo.com

Vine

This **app** doesn't exist any more (it ceased to operate in 2017) but it's here in case someone mentions it (and you can smugly say, 'But that doesn't exist any more'). When it *did* exist, Vine enabled users to film videos of up to six seconds long. These were mainly used to record (very) short music and comedy performances, slice-of-life vignettes, stop-motion animation and also as a promotional tool. In 1993 Dunkin' Donuts became the first company to use a single Vine video as an entire TV commercial.

Vintage

In the case of fashion, unlike **retro** style, which refers to new clothes, shoes and accessories that are inspired by old designs, the word vintage refers to items that are genuinely old but which have become very fashionable again. Examples of vintage clothing include American varsity jackets from the 1950s, tie-dye shirts from the 1960s or jumpsuits from the 1970s. You could walk down the street in these nowadays and people would think you're **on-trend**, rather than an idiot.

NB The jury is still out on flares and padded shoulders.

Viral

Social media makes it quick and easy to **share** something with people you know. That particular thing could be a photo, a video, animation, an article, **tweet** or **meme**. It could be something you find really funny or sad. It could also be something you're incensed about; something so annoying or shocking that you just have to let others see it, so they can share your anger or disgust.

Sometimes the 'thing' you share can catch the imagination and attention of the masses, taking on a life of its own and ending up being shared hundreds of thousands, or even millions, of times and it enters the public consciousness.

In this case, it's said that the photo, animation, meme, etc. has 'gone viral' – multiplying over and over and over and over again like a medical virus.

Why some things go viral and others don't is a mystery. Sometimes it's down to being in the right place at the right time; a few significant shares to the right audience can trigger a ripple effect and a whole **online** frenzy. In other cases, celebrities can **post** something provocative or absolutely newsworthy (well, to fans anyway) and that content can go viral within moments. Examples include the 2007 **YouTube** video 'Charlie Bit My Finger', which is now up to 860 million views. More recently, Kylie Jenner's **Instagram** photo of her new baby, announcing her name (Stormi Webster, if you're remotely interested), received 17.8 million likes; Kylie is the half-sister of Kim **Kardashian**.

But you knew that.

See also: **viral marketing**

Viral marketing

Remember back in 2014 seeing videos of people tipping buckets of ice-cold water over their heads for charity – then nominating the next person to take part? Maybe you did it too and **posted** the clip on **Facebook** or **Instagram**. That's an example – and a hugely successful one – of viral marketing; it raised $220 million worldwide for the condition ALS (also known as motor neurone disease). In essence, viral marketing is a technique where a promotional message (usually, but not exclusively, a short video) is delivered by a combination of word-of-mouth and **social media** (it's been described as the Holy Grail of marketing).

To encourage people to circulate/repost the video, it needs to be entertaining or provocative (or both); in essence it needs to give the audience a good experience. In a lot of cases the video isn't an advertisement as such, but something relevant to the brand. A good example is the film of Austrian skydiver Felix Baumgartner, who performed a freefall jump from 24 miles above the earth in 2012. Sponsored by Red Bull, with prominent branding on his pressure suit and at mission control, the video of his death-defying jump has been seen over 43 million times on **YouTube**.

See also: **viral**

Virtual reality

That perfect 10-yard putt to win the US Masters, precariously hanging on to an icy outcrop as you climb Everest, shooting down TIE-fighters in your X-wing fighter – all are achievable thanks to the wonders of virtual reality, or as it's more fashionably known, VR.

Promising a 'truly immersive experience', VR creates its artificial world by means of ultra-realistic computer-generated 3D images. The 'immersive' nature comes from wearing a VR headset or glasses and headphones that create the impression that you're in the actual environment. This realism is heightened by the fact that you can interact with your surroundings in a physical way (e.g. if you move, the image changes accordingly, as if you were really there).

Top quality VR is so realistic it can trick your senses into forgetting you're in this fake environment; this makes it as perfect for fighting aliens as it is for learning to fly a plane.

See also: **augmented reality**, **Oculus Rift**

Virus

In essence, a computer virus is a **program** designed with one purpose: to harm a computer system. It spreads by duplicating and attaching itself to your **files** and, just as biological viruses can vary in their seriousness (e.g. from chicken pox to Ebola), so too can their computer counterparts. If your computer or device is infected with a virus you might be lucky and just suffer from a slowdown in performance or finding it suddenly freezing. On the other hand, it could corrupt all of your data (your documents, photos, music, etc.) making it completely useless.

Like their medical counterparts, computer viruses spread quickly by duplicating. For example, infecting your **email** account and sending out virus-contaminated emails to everyone in your address book.

See also: **logic bomb**, **malware**

There are a number of ways to prevent your device or computer being infected:

* Use up-to-date anti-virus **software**. This will identify and stop infected **files** – and advise you when any have been detected.
* Don't open an email **attachment** unless you were expecting it and know the source.
* Only **download** files from reputable **websites**.
* Be wary about letting other people use their **flash drives** on your computer.
* Lastly, avoid using software obtained from unreliable sources (i.e. car boot sales and people in pubs).

To protect yourself from the effect of a virus it's good practice to **back-up** your data regularly. This means you can restore any damaged files from the uninfected back-up.

Vlog

Just as a **blog** is an **online** journal, think of a vlog as a video version (the word is a contraction of video log). And like blogs, they can be about any subject you damn well want – although relying on video **content** means they are more suited for demonstrating or reviewing stuff, e.g. beauty tips, interior design, fashion, lifestyle topics, healthy eating, exercise, video gaming, comedy, cars, photography, travel, etc.

Vlogs are hosted on a number of **social media platforms** but the most popular is **YouTube**. Others include **Vimeo**, **Facebook** and **Instagram**.

People who create vlogs are called vloggers, and the process of creating a vlog is known as vlogging. Vloggers can be huge stars in their own right, with millions of followers, earning millions of dollars from advertising and sponsorship. Twenty-seven-year-old Daniel Middleton (known as DanTDM) has an estimated 18 million subscribers to his YouTube channel, which has generated 12 billion views. He vlogs about the game Minecraft and in 2017 embarked on a world tour to talk about video games that included four sell-out nights at the Sydney Opera House.

To vlog, not only do you need to be able to talk to camera confidently and entertainingly about your chosen subject, but you need access to a high-quality video camera and microphone, good lighting and be able to edit your videos (or at least know someone who can do this for you). Some of the most successful vlogs are those with high production values. Oh yes. It also helps a lot if you're photogenic or, if not, then a bit quirky.

See also: **blog**, **Zoella**

From Wake word *to* World Wide Web

Wake word

This is the word (or words) used to activate your **digital assistant** and alert it to listen out for your request or command. The best-known wake words are '**Alexa**' and 'Hey **Siri**' but some devices allow you to customise the wake word to one of your own choosing (this will avoid confusing the device if you live in a household with someone called Alexa or, for that matter, Siri).

#WCW

Woman Crush Wednesday: an offshoot of **#MCM** but this time it's a **hashtag** used to **post** photos of women on **Twitter** or **Instagram** on Wednesdays (although some people use the hashtag to mean World Championship Wrestling, with obviously confusing consequences).

Web address

Another word (well, two words really), for **domain name**.

Webcam

Short for web camera (see what they did there?), this is a miniature digital camera connected to a computer. It can take stills or record or **stream** live video to another location via the internet. Most computers and **laptops** come with a webcam and microphone built in but if yours doesn't, you can buy a separate webcam and plug it in via your computer's **USB** port, or use one that works wirelessly via **Bluetooth**.

The two most common uses for webcams are for chatting via **Skype** and making a **sex tape**.

Webcast

A webcast is, quite simply, a broadcast in real-time over the internet. This could be from an internet-only broadcaster, a radio station or a TV channel that **simulcasts** a show over the airways and the internet, or a commercial organisation, which might be webcasting a presentation or an event such as a new product launch.
See also: **webinar**, **wedcast**

Some funeral homes provide a webcasting service. This allows mourners who cannot attend to pay their respects (and avoid having to stand in the cold and rain).

Webinar

Also known as web conferencing, a webinar is a live seminar delivered over the internet. Unlike a **webcast**, which is a one-way flow of information to the user, a webinar tends to be a more interactive experience with whoever's presenting it, including a live question-and-answer session or the ability to 'chat' to other participants. Webinars are often used for training sessions, workshops or lectures and are usually aimed at a more defined, invited audience who might pay to take part.
See also: **webcast**

Webmail

There are two types of **email** account. The first is the one provided by your **ISP**. Without going into too many technicalities this type of account **downloads** and stores emails directly onto your computer or **mobile device**. The downsides of this method are that you can only access email from that one device and you have limited storage for old emails (plus, if you change your ISP, you usually lose your whole email account).

Webmail, on the other hand, is the generic name given

to email services such as Outlook, Gmail and **Yahoo!** Mail that keep your emails in the **Cloud**. This means you can access them on any computer or device that's connected to the internet. The advantage is just that – as long as you have an internet connection you can read and send emails from anywhere in the world. The downsides are that if you're not connected, you can't access any emails at all, plus although webmail is provided free, it's paid for by advertising, so you might find various ads intrusively popping up every now and then.

Meh. What can you do?

The main advantage of having an ISP as your email provider is support. If anything goes wrong with your email, it's usually easier (in theory at least) to contact someone to resolve the issue.

Website

If the **World Wide Web** is a massive global reference library, then websites are all the books on its shelves. Except they're not books. They're a series of what's called 'web pages' that contain text, graphics, images, sound and video that are all viewable on your computer or any device connected to the internet. But apart from giving you almost instant access to information, what makes websites even better than books is that each web page has **hyperlinks** that will take you instantly to another page on that website – or to a different (but relevant) website entirely.

Yes, it's website not web site. Type 'What is a web site?' into **Google** and it will ask, 'Did you mean "What is a website?"'

And it should know.

Wedcast

You've decided to tie the knot in Cancun, Fiji or the Maldives but Aunty Vi suffers from acute lumbago and cousin Norman has a pathological fear of luggage, let alone flying. To prevent them and any one of your friends and relatives who can't make the trip missing out, many venues offer a wedcast, a cutesy name for a **webcast** whereby live video of the big day is **streamed** to their computers.

Wellness

Back in the day, if you weren't ill or infirm, you were well. Nowadays, though, there's the concept of wellness – that abstract feeling of being 'fulfilled' by being in a state of complete physical, mental and social wellbeing. Wellness as a notion has been around since the 1980s but was made 'official' by the foundation of the Global Wellness Institute (GWI) in 2014. This body has a mission to 'empower wellness worldwide' by research, initiatives and education.

To improve staff health and welfare, some large corporations or public bodies now employ Directors of Wellness; however, for everyone else who's concerned about achieving this state there is a multitude of companies cashing in on wellness, selling everything from crystal healing wands to magnetic shoe insoles.
See also: **hygge**, **mindfulness**

WhatsApp

Now owned by **Facebook**, this is currently the most popular **instant messaging app** with over 1.2 billion users worldwide. Using what's known as 'end-to-end **encryption**', it's also one of the most secure.
Visit: www.whatsapp.com

WiFi

These days we take for granted the ability to connect wirelessly to the internet, or to other devices such as a printer or a camera, but it's all thanks to this technology, which uses radio signals rather than cables to transmit and receive information.

Pronounced as 'Why-Fye' (never as 'Whiffy'), the word doesn't actually stand for anything. The name was coined in 1999 to allude to 'hi-fi', or high fidelity, a term that referred to high-quality music systems.

WikiLeaks

Nothing to do with **Wikipedia**, this is an international, non-profit organisation that's often in the news in relation to its founder Julian Assange, an Australian internet activist who had to hide out at Ecuador's embassy in London, where he was granted political asylum. Through its **website**, the organisation leaks sensitive and classified material in order to bring this information to the attention of the general public. Many of these leaks come from journalists and whistleblowers (the **content** is **uploaded** via an anonymous, untraceable and secure **hosting** service). Some of WikiLeak's best-known stories are those that reported US Army friendly-fire incidents, Guantanamo Bay operating procedures, the 'secret bibles' of Scientology and, very ironically, a British military manual that gave details of how to prevent leaks.

Wikipedia

A free **online** encyclopaedia with **content** created (and edited) by a worldwide team of volunteers known as Wikipedians. Wikipedia is the largest and most popular general reference work on the internet and is usually named as one of the most popular **websites**. The site currently offers 40 million articles in 299 different languages. Anyone can create or edit articles, so critics point out that the entries are unreliable – or vulnerable to devious editing (i.e. individuals, organisations or corporations can remove anything they see as negative – or, alternatively, insert negative comments into the entry of a rival).

Visit: www.wikipedia.org

'Wiki' is a Hawaiian word for quick. It reflects the fact that the website's articles include a multitude of links to quickly guide the user to related pages with even more information.

Windows

The **operating system** created by **Microsoft** and first released in 1985, which, in computing terms, is the equivalent of the Jurassic period. There are many versions of Windows, designed to work with a specific type of computer, e.g. desktop or **laptop** computers, mobile devices, **smartphones** and even the Xbox games console.

Word

To give it its full name, **Microsoft** Word is a very long-established and omnipresent word processor **program** designed to create and edit documents. Launched in 1983, it became the default product in commercial organisations worldwide; however, its dominance is being threatened by a new kid on the block – **Google** Docs. This web-based **software** from Google (duh!) is compatible with Word documents, works across multiple **operating systems** and, importantly, it's free!

World Wide Web

Whereas the internet is the global **network** of computers all communicating with each other, the World Wide Web ('www' or 'web' for short) is the collection of **websites** found on this network. It was invented by British scientist Sir Tim Berners-Lee in 1989 and although the internet already existed at this point (mainly for scientists and engineers to share information), it was Sir Tim who created the first web **browser** and devised a way to link one document directly to another – the feature that makes websites so useful. The technology he proposed to do this is still in use today.

From Yahoo! *to* YouTube

Yahoo!

One of the few companies with an exclamation mark in its name, Yahoo! is a major web services company that provides, among other things, a **search engine** (Yahoo! Search), **email** (Yahoo! Mail) a general news **website** (Yahoo! News) and a business news website (Yahoo! Finance).

Yahoo! was co-founded in 1994 by electrical engineering graduates Jerry Yang and David Filo. Its origins were a website that was a directory of other websites called 'Jerry and David's Guide to the **World Wide Web**'.

Youthquake

This word goes back to the 1960s to describe how the young were behind great cultural change. At the time it was about how they were influencing music and fashion but nowadays it's more about politics and social change.

YouTube

Owned by **Google** since 2006, this is the largest video-**sharing website** in the world and the go-to place to watch videos on absolutely everything, from the instructional (how to change your car's oil), the entertaining (the official video for Abba's 'Dancing Queen'), the odd (how to make a miniature coat hanger from a paper clip) and the just plain bonkers (a dancing man wearing a horse mask cooking wild mushrooms). YouTube allows users to **upload**, view, rate, share and comment on videos – and report those they find unsuitable. Individuals and corporations can also have their own YouTube channel.

It's free to upload videos to YouTube; most come from individuals (it's the most popular site for **vlogs**) but some media corporations also offer their material, e.g. film trailers, promotional videos. After Google, YouTube is ranked as the second most popular website in the world. *Visit:* www.youtube.com

Ten things you may not know about YouTube

1 The first YouTube video was uploaded on 23 April 2005. 'Me at the Zoo' shows co-founder Jawed Karim at the San Diego Zoo. It's been viewed over 50 million times.
2 The first video to hit one million views was a Nike commercial featuring the football player Ronaldinho.
3 It took the video for 'Gangnam Style' by Psy five months to hit 1 billion views. In comparison, 'Despacito' by Luis Fonsi took just ninety-seven days.
4 More than 400 hours of videos are uploaded to YouTube each minute.
5 And 1 billion hours are watched every day.
6 The most popular branded YouTube channel is currently Lego (which has amassed over 6 billion views).
7 YouTube star Grumpy Cat earned more money in 2014 than Oscar-winning actress Gwyneth Paltrow.
8 In 2018, 95 per cent of the most watched videos were music videos.
9 The most searched-for tutorial video is 'How to kiss'. The second is 'How to tie a tie'.
10 YouTube plays an April Fool's joke each year; in 2009 when users selected videos on the 'recommended for you' section, the whole webpage layout – including the video – played upside down.

From Zip file *to* Zuckerberg, Mark

Zip file

Trying to send a large computer **file** by **email** can cause problems. It will either take ages and ages to send or, if it's too big, it just won't get sent at all. In cases such as this, the answer is to compress the file, making it smaller without losing any of the information it contains (don't worry; it's all done for you by an **algorithm**). This compression is called 'zipping' the file. The result is a 'zip file', which you can then click on to decompress – or 'unzip'. Zip files can be recognised because they have .zip after their name.

Zoella

Zoe Elizabeth Sugg is a British fashion and beauty **vlogger**, and a huge **YouTube** star best known by her username Zoella. Broadcasting to over 12 million subscribers, Zoella's most popular videos include 'My make-up collection and

storage', 'My everyday make-up routine', 'My boyfriend does my make-up', 'My brother does my make-up' and 'What's in my handbag?' If you think beauty vlogging is a vapid and pointless thing to do, then think again. It's said she earns around £50,000 per month via YouTube royalties and endorsement deals. In 2017 *Forbes* magazine named her one of the most powerful influencers in the world of beauty.

Zuckerberg, Mark

Co-founder, chairman and CEO of **Facebook**, which he created and launched from his dorm room at Harvard University in 2004. In 2010 he was named *Time* magazine's Person of the Year, while *Vanity Fair* magazine named him number one in its list of the top 100 'Most Influential People of the Information Age'. Zuckerberg has featured perennially in subsequent lists of the 100 wealthiest and most influential people in the world.

Ten things you may not know about Mark Zuckerberg

1 His salary is $1 per year (although he is said to be currently worth around $66 billion).

2 He dropped out of Harvard to concentrate on developing Facebook.

3 His first project was the Facemash **program**, which compared two photos of Harvard students and allowed users to vote on who was hot and who was not. The university deemed it 'inappropriate' and closed it down.

4 He suffers from red-green colour-blindness and sees the colour blue best. That's why blue dominates Facebook's colour scheme.

5 His Hungarian sheepdog Beast has 2.7 million Facebook **followers**.

6 He wears the same style of grey T-shirt almost every day because it saves him time in the morning worrying about what to wear.

7 He was offered a job by **Microsoft** before he'd even left high school (which he turned down).
8 He donated $100 million to help rescue the struggling Newark Public Schools system of New Jersey, and $25 million to combat the spread of the Ebola virus in West Africa.
9 He voiced his own character in an episode of *The Simpsons*, telling Lisa that she doesn't need to graduate from college to be successful.
10 One of his first business cards stated, 'I'm CEO, Bitch'.

Also by Mark Leigh from Robinson

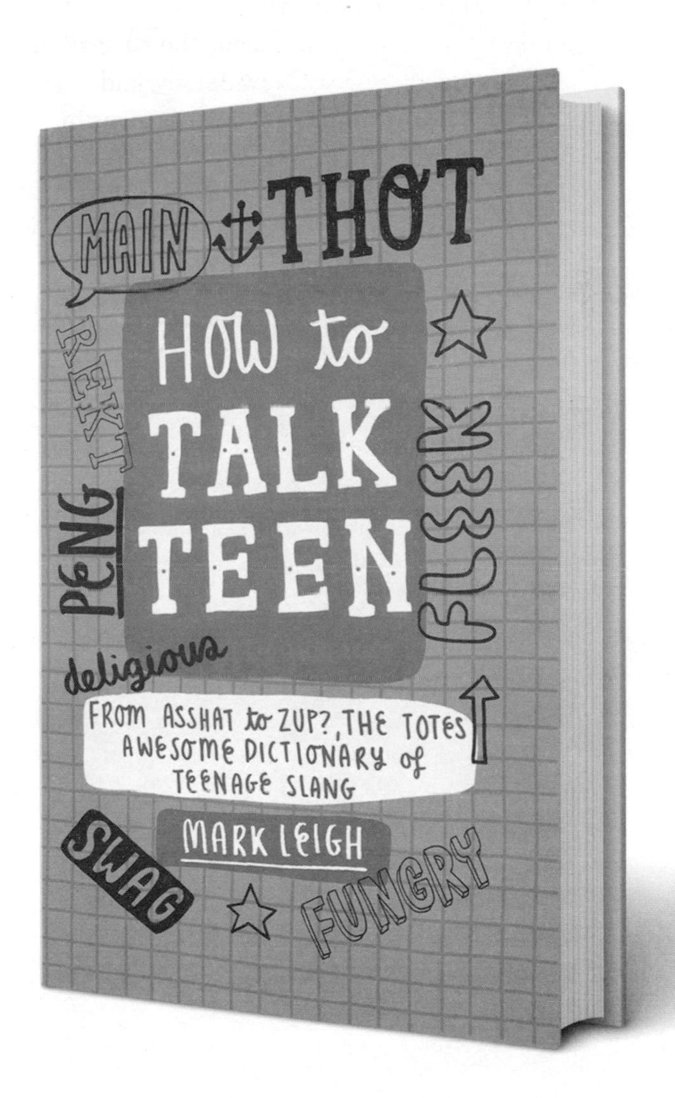